T0269285

Mathematical Biosciences Institute Lecture Series

The Mathematical Biosciences Institute (MBI) fosters innovation in the application of mathematical, statistical and computational methods in the resolution of significant problems in the biosciences, and encourages the development of new areas in the mathematical sciences motivated by important questions in the biosciences. To accomplish this mission, MBI holds many week-long research workshops each year, trains postdoctoral fellows, and sponsors a variety of educational programs.

The MBI lecture series are readable up to date introductions into exciting research areas that are inspired by annual programs at the MBI. The purpose is to provide curricular materials that illustrate the applications of the mathematical sciences to the life sciences. The collections are organized as independent volumes, each one suitable for use as a module in standard graduate courses in the mathematical sciences and written in a style accessible to researchers, professionals, and graduate students in the mathematical and biological sciences. The MBI lectures can also serve as an introduction for researchers to recent and emerging subject areas in the mathematical biosciences.

Marty Golubitsky, Michael Reed
Mathematical Biosciences institute

More information about this series at http://www.springer.com/series/13083

Mathematical Biosciences Institute Lecture Series
Volume 1: Stochastics in Biological Systems

Stochasticity is fundamental to biological systems. While in many situations the system can be viewed as a large number of similar agents interacting in a homogeneously mixing environment so the dynamics are captured well by ordinary differential equations or other deterministic models. In many more situations, the system can be driven by a small number of agents or strongly influenced by an environment fluctuating in space or time. Stochastic fluctuations are critical in the initial stages of an epidemic; a small number of molecules may determine the direction of cellular processes; changing climate may alter the balance among competing populations. Spatial models may be required when agents are distributed in space and interactions between agents form a network. Systems evolve to become more robust or co-evolve in response to competitive or host-pathogen interactions. Consequently, models must allow agents to change and interact in complex ways. Stochasticity increases the complexity of models in some ways, but may smooth and simplify in others.

Volume 1 provides a series of lectures by internationally well-known authors based on the year on Stochastics in biological systems which took place at the MBI in 2011–2012.

Michael Reed, Richard Durrett
Editors

Mathematical Biosciences Institute Lecture Series
Volume 1: Stochastics in Biological Systems

Stochastic Population and Epidemic Models
Linda S. Allen

Stochastic Analysis of Biochemical Systems
David Anderson; Thomas G. Kurtz

Branching Process Models of Cancer
Richard Durrett

Stochastic Neuron Models
Pricilla E. Greenwood; Lawrence M. Ward

The Mathematics of Intracellular Transport
Scott McKinley; Peter Kramer

Stochastic Models for Structured Populations
Sylvie Méléard; Vincent Bansaye

Probabilistic Models of Population Evolution
Etienne Pardoux

Correlations from Coupled Enzymatic Processing
Ruth Williams

Étienne Pardoux

Probabilistic Models
of Population Evolution

Scaling Limits, Genealogies and Interactions

 Springer

Étienne Pardoux
Institut de Mathématiques de Marseille
Aix-Marseille Université
Marseilles, France

ISSN 2364-2297   ISSN 2364-2300   (electronic)
Mathematical Biosciences Institute Lecture Series
ISBN 978-3-319-30326-0   ISBN 978-3-319-30328-4   (eBook)
DOI 10.1007/978-3-319-30328-4

Library of Congress Control Number: 2016937337

Mathematics Subject Classification (2010): 60J80, 60J85, 92D25, 60F17, 60H10

Printed on acid-free paper

Copublished with the Mathematical Biosciences Institute, Columbus, OH, USA

This Springer imprint is published by Springer Nature
The registered company is Springer International Publishing AG Switzerland

Contents

Chapter 1
Introduction

The aim of these notes is to describe several probabilistic models of population dynamics.

We start with the classical Bienaymé–Galton–Watson branching process, both in discrete and in continuous time. The second step consists of describing continuous state branching processes (in short CSBPs) and showing how such a process can be obtained as a limit of properly rescaled BGW processes, as the population size tends to infinity. Note that we restrict ourselves here to continuous (i.e., Feller type) CSBPs.

The next topic consists in describing the genealogies of the population. A population is not described only by the time evolution of its size, i.e., not only by a $\mathbb{Z}_+$-valued function of time. The tree (or the forest of trees) coding the genealogy and the family relations among the individuals contain much more information. The time evolution of the population size can be recovered from the genealogical tree, but the converse is not true. One of our aims in this part of the notes is to describe the genealogical forest of trees corresponding to a CSBP. There is no way to draw a forest of trees in the conventional way that trees are drawn. However, following the work of Aldous [2], trees can be described by their contour process (see Figure 5.2), and the law of the contour of the genealogical forest of trees corresponding to a Feller type CSBP can be characterized as that of a reflected Brownian motion (at least in the critical case). This is one way to interpret the second Ray–Knight theorem, of which several generalizations are exposed in these notes.

Next, and this is probably the main originality of these notes, we want to describe the evolution of a population where the birth or death rates of the various individuals are affected by the size of the population. This may account for competition for rare resources. In that case, the interaction would increase the death rate of the individuals. However, the increase in population size could very well increase the birth rate (or decrease the death rate but we shall not consider this alternative). This is the so-called Allee effect. In all these cases, the population process is of course no longer a branching process, due to the interactions.

© Springer International Publishing Switzerland 2016
É. Pardoux, *Probabilistic Models of Population Evolution*, Mathematical Biosciences Institute Lecture Series 1.6, DOI 10.1007/978-3-319-30328-4_1

In order to describe the evolution of those interacting collections of individuals jointly for all ancestral population sizes, we need to introduce a pecking order (which we arbitrarily choose from left to right), which is transmitted by each individual to her descendants. Any individual interacts at time t with all individuals alive at that time, and sitting on her left, not with those on her right. Thanks to this, the interaction with the rest of the population felt by one of the descendants of ancestor k at some time t is the same, whether the number of ancestors at time 0 is $k, k+1, \ldots$, as it should be.

This pecking order is also crucial for the description of the genealogical forest of trees in the population with interactions. This forest of trees is described both in the case of a finite population and in the rescaled infinite population limit, the last one being obtained as the limit of the former as the size of the ancestral population tends to infinity.

Consider the extinction time and the total length (in the continuous set-up we say "total mass") of the forest of trees. Without interaction, it is easy to prove that these quantities tend to infinity as the number of ancestors (in the continuous case as the "mass" of ancestors) at time 0 tends to infinity. We will prove that if the competition is strong enough, those limits may be finite, and even have some finite exponential moments.

These notes are organized as follows. Chapter 2 describes the branching processes. Section 2.1 presents the classical discrete time branching processes and the Bienaymé–Galton–Watson processes. Section 2.2 describes the continuous time branching processes. Chapter 3 presents convergence results of the rescaled versions of the above processes, as the initial population size tends to infinity. Chapter 4 presents the Feller type CSBPs. We start with the Dawson–Li stochastic differential equation for these CSBPs (which in our case is a clever way of writing the Feller SDE jointly for all ancestral masses), then compute the Laplace functional of those processes, and analyze in more detail the number of ancestral individuals who contribute to the population at time t. Finally, we study the Dawson–Li SDE for continuous population with interaction. Chapter 5 describes the genealogical forest of trees corresponding to the above models, and shows that the genealogy of the CSBP can be obtained as a limit of the genealogies of approximating continuous time binary branching processes. Chapter 6 presents the models for finite populations with interaction and its genealogical forest of trees. We also consider the effect of the interaction on the extinction time and total length of those trees, as the ancestral population tends to infinity. Chapter 7 shows how to take the limit in the finite population process with interaction, and in its genealogical forest of trees, as the ancestral population size tends to infinity. Chapter 8 describes the genealogical forest of trees of the continuous state population process with interaction, and the effect of the interaction on the extinction time and total mass of those trees, as the ancestral population mass tends to infinity. Finally the Appendix collects most of the technical tools and results which are used in these notes. Only a few of those results are proved. Most of them are just stated. They should help the reader who may not remember some of the technical notions which are used. For those who need a more complete introduction, references are given to more complete texts.

These notes are intended mainly for readers with some basic knowledge of stochastic processes and stochastic calculus.

It is my pleasure to thank my collaborators and students: Mamadou Ba, Vi Le, and Anton Wakolbinger, with whom some of the results exposed here have been obtained.

Chapter 2
Branching Processes

A branching process is a $\mathbb{Z}_+$-valued process $\{X_t,\ t \in \mathbb{Z}_+ \text{ or } t \in \mathbb{R}_+\}$ which is such that for each t, $\{X_{t+s},\ s > 0\}$ is the sum of X_t independent copies of $\{X_s,\ s > 0\}$, where the latter starts from $X_0 = 1$. This type of process models the evolution of a population where the progenies of various individuals are i.i.d., i.e. there is no interaction between various contemporaneous individuals, whose fertility and lifetime have the same law. Models with interaction between the individuals will be studied below, starting with chapter 6.

We first study discrete time branching processes in section 2.1, and then continuous time Markov branching processes in section 2.2, distinguishing the general case from the case of binary branching.

2.1 Discrete Time Bienaymé–Galton–Watson Processes

Consider a Bienaymé–Galton–Watson process, i.e., a process $\{X_n,\ n \geq 0\}$ with values in $\mathbb{N}$ (n denotes the generation and X_n the size of the n-th generation of the population) such that

$$X_{n+1} = \sum_{k=1}^{X_n} \xi_{n,k},$$

where $\{\xi_{n,k},\ n \geq 0, k \geq 1\}$ are i.i.d. r. v.'s with as joint law that of ξ whose generating function $f(s) = \mathbb{E}[s^\xi] = \sum_{k \geq 0} s^k \mathbb{P}(\xi = k)$ satisfies

$$\mu := \mathbb{E}[\xi] = f'(1) = 1 + r, \text{ and } 0 < f(0) = \mathbb{P}(\xi = 0) < 1.$$

We call f the probability generating function (p. g. f. in short) of the Bienaymé–Galton–Watson process $\{X_n,\ n \geq 0\}$. In order to exclude trivial situations, we assume that $\mathbb{P}(\xi = 0) = f(0) > 0$ and that $\mathbb{P}(\xi > 1) > 0$. This last condition implies that $s \to f(s)$, which is increasing on $[0,1]$, is a strictly convex function.

© Springer International Publishing Switzerland 2016
É. Pardoux, *Probabilistic Models of Population Evolution*, Mathematical
Biosciences Institute Lecture Series 1.6, DOI 10.1007/978-3-319-30328-4_2

The process is said to be *subcritical* if $\mu < 1$ $(r < 0)$, *critical* if $\mu = 1$ $(r = 0)$, and *supercritical* if $\mu > 1$ $(r > 0)$.

First note that the process $\{X_n,\ n \geq 0\}$ is a Markov chain, which possesses the so-called branching property, which we now formulate. For $x \in \mathbb{N}$, let $\mathbb{P}_x$ denote the law of the Markov process $\{X_n,\ n \geq 0\}$ starting from $X_0 = x$. The law of $\{X_n,\ n \geq 0\}$ under $\mathbb{P}_{x+y}$ is the same as that of the sum of two independent copies of $\{X_n,\ n \geq 0\}$, one having the law $\mathbb{P}_x$, the other the law $\mathbb{P}_y$. The branching property follows from the assumption that the $\xi_{n,k}$'s are mutually independent (and have the same law), which means that there is no interaction between the individuals. We shall introduce interactions further below, starting with chapter 6.

We next define

$$T = \inf\{k > 0; X_k = 0\},$$

which is the time of extinction. We first recall

Proposition 1. *Assume that $X_0 = 1$. Then the probability of extinction $\mathbb{P}(T < \infty)$ is one in the subcritical and the critical cases, and it is the unique root $q < 1$ of the equation $f(s) = s$ in the supercritical case.*

PROOF: Let $f^{\circ n}(s) := f \circ \cdots \circ f(s)$, where f has been composed n times with itself. It is easy to check that $f^{\circ n}$ is the generating function of the r. v. X_n; in case $X_0 = 1$.

On the other hand, clearly $\{T \leq n\} = \{X_n = 0\}$. Consequently

$$\mathbb{P}(T < \infty) = \lim_n \mathbb{P}(T \leq n)$$
$$= \lim_n \mathbb{P}(X_n = 0)$$
$$= \lim_n f^{\circ n}(0).$$

Now the function $s \to f(s)$ is continuous, increasing, and strictly convex, starts from $\mathbb{P}(\xi = 0) > 0$ at $s = 0$, and ends at 1 at $s = 1$. If $\mu = f'(1) \leq 1$, then $\lim_n f^{\circ n}(0) = 1$. If, however, $f'(1) = 1 + r > 1$, then there exists a unique $0 < q < 1$ such that $f(q) = q$, and it is easily seen that $q = \lim_n f^{\circ n}(0)$. These last arguments are easy to understand by looking at Figure 2.1 below. □

Note that the state 0 is absorbing for the Markov chain $\{X_n,\ n \geq 0\}$, and it is accessible from each state. It is then easy to deduce that all other states are transient, hence either $X_n \to 0$ or $X_n \to \infty$, as $n \to \infty$. In other words, the population tends to infinity a.s. on the set $\{T = \infty\}$. Define $\sigma^2 = \mathrm{Var}(\xi)$, which is assumed to be finite. We have

Lemma 1.

$$\mathbb{E}X_n = \mu^n \mathbb{E}X_0$$

$$\mathbb{E}[X_n^2] = \frac{\mu^{2n} - \mu^n}{\mu^2 - \mu}\sigma^2 \mathbb{E}X_0 + \mu^{2n}\mathbb{E}(X_0^2).$$

(In the case $\mu = 1$, the factor of $\sigma^2 \mathbb{E}X_0$ should be replaced by n, which is its limit as $\mu \to 1$.)

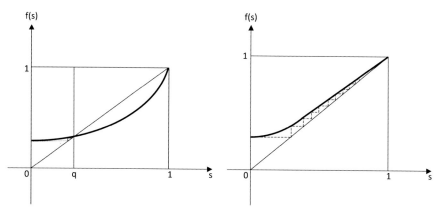

Fig. 2.1 Typical graphs of f when $\mu > 1$ (left) and $\mu \le 1$ (right).

PROOF: We have

$$\mathbb{E}X_n = \mathbb{E}\left[\mathbb{E}\left[\sum_{k=1}^{X_{n-1}} \xi_{n-1,k}\Big|X_{n-1}\right]\right]$$
$$= \mu\mathbb{E}X_{n-1}$$
$$= \mu^n\mathbb{E}X_0,$$

and

$$\mathbb{E}[X_n^2] = \mathbb{E}\left[\mathbb{E}\left[\left(\sum_{k=1}^{X_{n-1}} \xi_{n-1,k}\right)^2\Big|X_{n-1}\right]\right]$$
$$= \mu^2\mathbb{E}[X_{n-1}(X_{n-1}-1)] + (\sigma^2+\mu^2)\mathbb{E}X_{n-1}$$
$$= \mu^2\mathbb{E}[X_{n-1}^2] + \sigma^2\mathbb{E}X_{n-1}$$
$$= \mu^2\mathbb{E}[X_{n-1}^2] + \sigma^2\mu^{n-1}\mathbb{E}X_0.$$

Consequently $a_n := \mu^{-2n}\mathbb{E}[X_n^2]$ satisfies

$$a_n = a_{n-1} + \sigma^2\mu^{-(n+1)}\mathbb{E}X_0$$
$$= a_0 + \sigma^2\mathbb{E}X_0\sum_{k=1}^{n}\mu^{-(k+1)}.$$

The result follows. $\qquad\square$

Let now X_n^* denote the number of individuals in generation n with an infinite line of descent. Under $\mathbb{P}_1$, $\{T = \infty\} = \{X_0^* = 1\}$. ξ denoting an r. v. whose law is that of the number of offsprings of each individual, let $\xi^* \le \xi$ denote the number of those offsprings with an infinite line of descent. Let $\overline{q} := 1 - q = \mathbb{P}_1(T = \infty)$. We have in the supercritical case

Proposition 2. *Assume that $X_0 = 1$.*

1. *Conditionally upon $\{T = \infty\}$, $\{X_n^*, \, n \geq 0\}$ is again a Bienaymé–Galton–Watson process, whose p.g.f. is given by*

$$f^*(s) = [f(q + \bar{q}s) - q]/\bar{q}.$$

2. *Conditionally upon $\{T < \infty\}$, the law of $\{X_n, \, n \geq 0\}$ is that of a Bienaymé–Galton–Watson process, whose p.g.f. is given by*

$$\tilde{f}(s) = f(qs)/q.$$

3. *For all $0 \leq s, t \leq 1$,*

$$\mathbb{E}\left[s^{\xi - \xi^*} t^{\xi^*} \right] = f(qs + \bar{q}t)$$

$$\mathbb{E}\left[s^{X_n - X_n^*} t^{X_n^*} \right] = f^{\circ n}(qs + \bar{q}t).$$

4. *Conditionally upon $\{T = \infty\}$, the law of $\{X_n, \, n \geq 0\}$ is that of $\{X_n^*, \, n \geq 0\}$ to which we add individuals with finite line of descent, by attaching to each individual of the tree of the X_n^*'s N independent copies of a Bienaymé–Galton–Watson tree with p. g. f. $\tilde{f}$, where*

$$\mathbb{E}[s^N | X^*] = \frac{D^n f(qs)}{D^n f(q)},$$

where $D^n f$ denotes the n-th derivative of f, and n is the number of daughters of the considered individual in the tree X^*.

PROOF: Let us first prove the first part of *3*. Consider on the same probability space mutually independent r. v.'s $\{\xi, Y_i, \, i \geq 1\}$, where the law of ξ is given as above, and $\mathbb{P}(Y_i = 1) = \bar{q} = 1 - \mathbb{P}(Y_i = 0)$, $\forall i \geq 1$. Note that $\bar{q}$ is the probability that any given individual has an infinite line of descent, so that the joint law of $(\xi - \xi^*, \xi^*)$ is that of

$$\left(\sum_{i=1}^{\xi} (1 - Y_i), \sum_{i=1}^{\xi} Y_i \right).$$

$$
\begin{aligned}
\mathbb{E}\left[s^{\xi - \xi^*} t^{\xi^*} \right] &= \mathbb{E}\left[s^{\sum_{i=1}^{\xi}(1-Y_i)} t^{\sum_{i=1}^{\xi} Y_i} \right] \\
&= \mathbb{E}\left[\mathbb{E}[s^{1-Y_1} t^{Y_1}]^{\xi} \right] \\
&= \mathbb{E}[(qs + \bar{q}t)^{\xi}] \\
&= f(qs + \bar{q}t).
\end{aligned}
$$

A similar computation yields the second statement in *3*. Indeed

$$\mathbb{E}\left[s^{X_n - X_n^*} t^{X_n^*}\right] = \mathbb{E}\left[\mathbb{E}\left(s^{X_n - X_n^*} t^{X_n^*} | X_{n-1}\right)\right]$$
$$= \mathbb{E}\left[\left(\mathbb{E}\left[s^{\xi - \xi^*} t^{\xi^*}\right]\right)^{X_{n-1}}\right]$$
$$= f^{\circ(n-1)}(f(qs + \overline{q}t))$$

We next prove *1* as follows:

$$\mathbb{E}\left(t^{\xi^*} | \xi^* > 0\right) = \frac{\mathbb{E}(1^{\xi - \xi^*} t^{\xi^*}; \xi^* > 0)}{\mathbb{P}(\xi^* > 0)}$$
$$= \frac{\mathbb{E}(1^{\xi - \xi^*} t^{\xi^*}) - \mathbb{E}(1^{\xi - \xi^*} t^{\xi^*}; \xi^* = 0)}{\mathbb{P}(\xi^* > 0)}$$
$$= \frac{f(q + \overline{q}t) - q}{\overline{q}}.$$

We now prove *2*. It suffices to compute

$$\mathbb{E}\left(s^{\xi} | \xi^* = 0\right) = \mathbb{E}\left(s^{\xi - \xi^*} | \xi^* = 0\right)$$
$$= \frac{f(sq + 0\overline{q})}{q}.$$

Finally we prove *4*. All we have to show is that

$$\mathbb{E}[s^{\xi - \xi^*} | \xi^* = n] = \frac{D^n f(qs)}{D^n f(q)}.$$

This follows from the two following identities:

$$n! \mathbb{E}[s^{\xi - \xi^*}; \xi^* = n] = \overline{q}^n D^n f(qs + \overline{q}t)|_{t=0}$$
$$= \overline{q}^n D^n f(qs),$$
$$n! \mathbb{P}(\xi^* = n) = \overline{q}^n D^n f(qs + \overline{q}t)|_{s=1, t=0}$$
$$= \overline{q}^n D^n f(q).$$

$\square$

2.2 Continuous Time Markov Branching Process

2.2.1 The General Case

Consider a continuous time $\mathbb{N}$-valued branching process $X = \{X_t^k, \ t \geq 0, k \in \mathbb{N}\}$, where t denotes time, and k is the number of ancestors at time 0. Such a process is a

Bienaymé–Galton–Watson process in which to each individual is attached a random vector describing her lifetime and her number of offsprings. We assume that those random vectors are i.i.d. The rate of reproduction is governed by a finite measure μ on $\mathbb{N}$, satisfying $\mu(1) = 0$. More precisely, each individual lives for an exponential time with parameter $\mu(\mathbb{N})$, and is replaced by a random number of children according to the probability $\mu(\mathbb{N})^{-1}\mu$. Hence the dynamics of the continuous time jump Markov process X is entirely characterized by the measure μ. We have the

Proposition 3. *The generating function of the process X is given by*

$$\mathbb{E}\left(s^{X_t^k}\right) = \psi_t(s)^k, \; s \in [0,1], \; k \in \mathbb{N},$$

where

$$\frac{\partial \psi_t(s)}{\partial t} = \Phi(\psi_t(s)), \quad \psi_0(s) = s,$$

and the function Φ is defined by

$$\Phi(s) = \sum_{n=0}^{\infty} (s^n - s)\mu(n)$$
$$= \lambda(h(s) - s), \; s \in [0,1],$$

where $\lambda = \mu(\mathbb{N})$ and h is the generating function of the probability measure $\lambda^{-1}\mu$.

PROOF: Note that the process X is a continuous time $\mathbb{N}$-valued jump Markov process, whose infinitesimal generator is given by

$$Q_{n,m} = \begin{cases} 0, & \text{if } m < n-1, \\ n\mu(m+1-n), & \text{if } m \geq n-1 \text{ and } m \neq n, \\ -n\mu(\mathbb{N}), & \text{if } m = n. \end{cases}$$

Define $f : \mathbb{N} \to [0,1]$ by $f(k) = s^k$, $s \in [0,1]$. Then $\psi_t(s) = P_t f(1)$. It follows from the backward Kolmogorov equation for the process X (see, e.g., Theorem 3.2, Chapter 7 in [34]) that

$$\frac{dP_t f(1)}{dt} = (QP_t f)(1)$$
$$\frac{\partial \psi_t(s)}{\partial t} = \sum_{k=0}^{\infty} Q_{1,k} \psi_t(s)^k$$
$$= \sum_{k=0}^{\infty} \mu(k) \psi_t(s)^k - \psi_t(s) \sum_{k=0}^{\infty} \mu(k)$$
$$= \Phi(\psi_t(s)).$$

$\square$

The branching process X is called immortal if $\mu(0) = 0$.

2.2.2 The Binary Branching Case

We now consider the case where $\mu(k) = 0$, for all $k \notin \{0,2\}$. In this situation, there is another equivalent description of the model. Let $b = \mu(2)$, $d = \mu(0)$. Any individual lives for a duration which is exponential with parameter d, and during her life gives birth to new individuals (one at a time) according to a rate b Poisson process $P(t)$ (also called a Poisson counting process), that is an increasing $\mathbb{N}$-valued process with independent increments, which is such that for any $0 \leq s < t$, $P(t) - P(s)$ (which counts a random number of points in the interval $(s,t]$, see below section 5.2) follows the Poisson distribution with parameter $b(t-s)$.

The process X is a birth and death process, which for $i \geq 1$ jumps from i to $i+1$ at rate ib, and from i to $i-1$ at rate id, and is absorbed at 0.

Chapter 3
Convergence to a Continuous State Branching Process

If one wants to understand the evolution of a large population (e.g., in order to study its extinction time), it may be preferable to consider the limit, as the population size tends to infinity, of the rescaled $\mathbb{Z}_+$-valued branching process. The limit, which is $\mathbb{R}_+$-valued, inherits a branching property, that of the so-called continuous state branching process (in short CSBP). The formal statement of the CSBP property, which is very similar to the formulation of the branching property as stated before Proposition 1 in Chapter 2, will be given at the start of Chapter 4. In the present chapter, we will show convergence results of rescaled branching processes towards the solution of a Feller SDE. Note that we consider only convergence towards CSBPs with continuous trajectories, hence towards a Feller diffusion. More general CSBPs will be alluded to below in Remark 2 of Chapter 4. For the convergence of branching processes towards those general CSBPs, we refer to Duquesne, Le Gall [17] and Grimvall [20].

3.1 Convergence of Discrete Time Branching Processes

Let $x > 0$ be a given real number. To each integer N, we associate a Bienaymé–Galton–Watson process $\{X_n^{N,x}, \ n \geq 0\}$ starting from $X_0^{N,x} = [Nx]$. We now define the rescaled continuous time process

$$Z_t^{N,x} := N^{-1} X_{[Nt]}^{N,x}.$$

We shall let the p. g. f. of the Bienaymé–Galton–Watson process depend upon N in such a way that

$$\mathbb{E}[\xi^N] = f_N'(1) = 1 + \frac{\gamma_N}{N},$$
$$\text{Var}[\xi^N] = \sigma_N^2,$$

© Springer International Publishing Switzerland 2016
É. Pardoux, *Probabilistic Models of Population Evolution*, Mathematical Biosciences Institute Lecture Series 1.6, DOI 10.1007/978-3-319-30328-4_3

where as $N \to \infty$,

$$\gamma_N \to \gamma \in \mathbb{R}, \quad \sigma_N \to \sigma. \text{[1]} \tag{3.1}$$

We assume in addition the following Lindeberg type condition

$$\mathbb{E}\left[|\xi^N|^2; \xi^N \geq a\sqrt{N}\right] \to 0 \text{ as } N \to \infty, \text{ for all } a > 0, \tag{3.2}$$

where we have used the notation $\mathbb{E}[X;A] = \mathbb{E}[X\mathbf{1}_A]$.

We denote by $\xi_j^{N,i}$ the number of offsprings of the j-th individual from generation i. $(\xi_j^{N,i})_{i \geq 0, \, j \geq 1}$ are i.i.d. with the above law. We have

$$Z_t^{N,x} = Z_{\frac{[Nt]}{N}}^{N,x}$$

$$= \frac{[Nx]}{N} + \frac{1}{N}\sum_{i=0}^{[Nt]-1}\sum_{j=1}^{NZ_{i\Delta t}^{N,x}}(\xi_j^{N,i} - 1).$$

Hence adding and subtracting γ_N/N in each term of the last double sum, we deduce that

$$Z_t^{N,x} = \frac{[Nx]}{N} + \gamma_N \int_0^{\frac{[Nt]}{N}} Z_s^N ds + M_t^N, \tag{3.3}$$

where $M_t^N = M_{\frac{[Nt]}{N}}^N$, with $M_{k\Delta t}^N = \tilde{M}_k^N$, and $\{\tilde{M}_k^N, k \geq 0\}$ is a discrete time martingale given by

$$\tilde{M}_k^N = \frac{1}{N}\sum_{i=0}^{k-1}\sum_{j=1}^{NZ_{i\Delta t}^N}\left[\xi_j^{N,i} - \left(1 + \frac{\gamma_N}{N}\right)\right].$$

It is shown in [20] that under conditions (3.1) and (3.2)[2] (for the definition of the space $D([0,+\infty); \mathbb{R}_+)$, see section A.7 below),

Proposition 4. $Z^{N,x} \Rightarrow Z^x$ in $D([0,+\infty); \mathbb{R}_+)$ equipped with the Skohorod topology, where $\{Z_t^x, t \geq 0\}$ solves the SDE

$$dZ_t^x = \gamma Z_t^x dt + \sigma\sqrt{Z_t^x}dB_t, \quad t \geq 0, \quad Z_0^x = x. \tag{3.4}$$

The proof in [20] is based on Laplace transform calculations. We will now give a proof based on martingale arguments. We deduce easily from (3.3)

Lemma 2. For any $N \geq 1$, $t > 0$,

$$\mathbb{E}[Z_t^{N,x}] \leq x\exp(\gamma_N t).$$

---

[1] The particular choice $\sigma = 2$ would introduce simplifications in many formulas of Chapters 5, 6, 7, and 8 below.

[2] In fact the result is proved in [20] under the slightly weaker assumption $\mathbb{E}\left[|\xi^N|^2; \xi^N \geq aN\right] \to 0$ as $N \to \infty$, for all $a > 0$.

PROOF: Taking the expectation in (3.3), we obtain the inequality

$$\mathbb{E}[Z_t^{N,x}] \leq x + \gamma_N \int_0^t \mathbb{E}[Z_s^{N,x}] ds,$$

from which the result follows, thanks to Gronwall's Lemma. □

Note that M_t^N is not a continuous time martingale, but $M_{k\Delta t}^N = \tilde{M}_k^N$ is a discrete time martingale. Let

$$[\tilde{M}^N]_k = \sum_{j=0}^{k-1} (\tilde{M}_{j+1}^N - \tilde{M}_j^N)^2.$$

It is easily shown that $\{(\tilde{M}_k^N)^2 - [\tilde{M}^N]_k, \ k \geq 1\}$ is a martingale. Moreover, from the fact that with $\mathscr{G}_i^N = \sigma\{Z_{j\Delta t}^{N,x}, \ j \leq i\}$,

$$\mathbb{E}\left[(\tilde{M}_{j+1}^N - \tilde{M}_j^N)^2 | \mathscr{G}_i^N \right] = \sigma_N^2 Z_{j\Delta t}^{N,x} \Delta t,$$

we deduce that

$$(\tilde{M}_k^N)^2 - \langle \tilde{M}^N \rangle_k \ \text{ is a martingale,} \tag{3.5}$$

where

$$\langle \tilde{M}^N \rangle_k = \sigma_N^2 \int_0^{k\Delta t} Z_s^N ds,$$

We can now prove the

Lemma 3. *For any $T > 0$,*

$$\sup_{N \geq 1} \mathbb{E} \left(\sup_{0 \leq t \leq T} Z_t^{N,x} \right) < \infty.$$

PROOF: In view of (3.3) and Lemma 2, it suffices to estimate

$$\left[\mathbb{E} \left(\sup_{0 \leq k \leq NT} |\tilde{M}_k^N| \right) \right]^2 \leq \mathbb{E} \left(\sup_{0 \leq k \leq NT} |\tilde{M}_k^N|^2 \right)$$

$$\leq 4\mathbb{E} \left(|\tilde{M}_{[NT]}^N|^2 \right)$$

$$= 4\mathbb{E} \langle \tilde{M}^N \rangle_{[NT]}$$

$$\leq 4\sigma_N^2 \int_0^T \mathbb{E}(Z_s^{N,x}) ds$$

$$\leq 4x\sigma_N^2 \frac{\exp[\gamma_N T] - 1}{\gamma_N},$$

where we have used Doob's inequality on the second line, (3.5) on the third, and with the understanding that the last ratio equals T if $\gamma_N = 0$. □

Remark 1. It is not very hard to show that

$$\mathbb{E}\left[(Z_t^{N,x})^2\right] \leq \left(x^2 + x\sigma_N^2 \frac{e^{\gamma_N t} - 1}{\gamma_N}\right) e^{\gamma_N(2 + \frac{\gamma_N}{N})t}$$

$$\mathbb{E}\left[\sup_{0 \leq s \leq t} (Z_s^{N,x})^2\right] \leq C(t)(x + x^2).$$

However, we do not need those estimates, and we leave their proof to the reader.

We can now proceed to the

PROOF OF PROPOSITION 4 Since M_t^N is not really a martingale, the arguments of Propositions 37 and 38 need to be slightly adapted. We omit the details of those adaptations. It follows from (3.3), (3.5), Lemma 3, Proposition 37 (see also Remark 14) and (3.1) that $\{Z_t^{N,x}, \ t \geq 0\}_{N \geq 1}$ is tight in $D([0, \infty); \mathbb{R}_+)$. In order to show that

$$Z_t^x = x + \gamma \int_0^t Z_s^x ds + M_t,$$

where M is a continuous martingale such that

$$\langle M \rangle_t = \sigma^2 \int_0^t Z_s^x ds,$$

see Proposition 38, it remains to prove that the last condition of Proposition 37 holds, namely that

Lemma 4. *For any $T > 0$, as $N \to \infty$,*

$$\sup_{0 \leq t \leq T} |Z_t^{N,x} - Z_{t^-}^{N,x}| \to 0 \ \text{in probability.}$$

If we admit for a moment this Lemma, it follows from the martingale representation Theorem 19 that there exists a standard Brownian motion $\{B_t, t \geq 0\}$ such that

$$Z_t^x = x + \gamma \int_0^t Z_s^x ds + \sigma \int_0^t \sqrt{Z_s^x} dB_s, t \geq 0.$$

It follows from Corollary 1 below that this SDE has a unique solution, hence the limiting law of $\{Z_t^x, t \geq 0\}$ is uniquely characterized, and the whole sequence $\{Z^{N,x}\}$ converges to Z^x as $N \to \infty$. $\qquad \square$

PROOF OF LEMMA 4 For any $\varepsilon' < \varepsilon$, provided N is large enough,

$$\mathbb{P}\left(\sup_{0 \leq t \leq T} |Z_t^{N,x} - Z_{t^-}^{N,x}| > \varepsilon\right) \leq \mathbb{P}\left(\sup_{i \leq NT} |\tilde{M}_i^N - \tilde{M}_{i-1}^N| > \varepsilon'\right)$$

Now define

$$\tilde{M}_k^{N,K} = \frac{1}{N} \sum_{i=0}^{k-1} \sum_{j=1}^{NZ_{i\Delta t}^{N,K}} \left[\xi_j^{N,i} - \left(1 + \frac{\gamma_N}{N}\right)\right],$$

where $Z_{i\Delta t}^{N,K} = Z_{i\Delta t}^N \wedge K$. It follows from Lemma 3 that for all $T > 0$,

$$\lim_{K \to \infty} \mathbb{P}(\tilde{M}_k^{N,K} = \tilde{M}_k^N \text{ for all } k \leq NT, \text{ all } N \geq 1) = 1.$$

It thus suffices to show that for each fixed $K > 0$, $\varepsilon > 0$,

$$\mathbb{P}\left(\sup_{i<NT} |\tilde{M}_{i+1}^{N,K} - \tilde{M}_i^{N,K}| > \varepsilon \right) \to 0,$$

as $N \to \infty$. But

$$|\tilde{M}_{i+1}^{N,K} - \tilde{M}_i^{N,K}| \leq \sqrt{\frac{K}{N}} |U_i^N|, \quad \text{where } U_i^N = \frac{1}{\sqrt{NZ_{i\Delta t}^{N,K}}} \sum_{j=1}^{NZ_{i\Delta t}^{N,K}} \bar{\xi}_j^{N,i},$$

with $\bar{\xi}_j^{N,i} = \xi_j^{N,i} - (1 + \gamma_N/N)$. We have with $\varepsilon_K = \varepsilon/\sqrt{K}$,

$$\mathbb{P}\left(\sup_{i \leq NT} |\tilde{M}_i^{N,K} - \tilde{M}_{i-1}^{N,K}| > \varepsilon \right) \leq \mathbb{P}\left(\bigcup_{i<NT} \left\{ |U_i^N| > \varepsilon_K \sqrt{N} \right\} \right)$$

$$= 1 - \mathbb{P}\left(\bigcap_{i<NT} \left\{ |U_i^N| \leq \varepsilon_K \sqrt{N} \right\} \right)$$

It is plain that

$$\mathbb{P}\left(|U_i^N| \leq \varepsilon_K \sqrt{N} \,\Big|\, \mathscr{G}_i^N \right) = 1 - \mathbb{P}\left(|U_i^N| > \varepsilon_K \sqrt{N} \,\Big|\, \mathscr{G}_i^N \right)$$

$$\geq 1 - \frac{C_{N,K}}{\varepsilon_K^2 N},$$

where

$$C_{N,K} = \sup_{0<z\leq K} \mathbb{E}\left(\frac{1}{[Nz]} \left| \sum_{j=1}^{[Nz]} \bar{\xi}_j^{N,i} \right|^2 ; \frac{1}{\sqrt{[Nz]}} \left| \sum_{j=1}^{[Nz]} \bar{\xi}_j^{N,i} \right| > \varepsilon_K \sqrt{N} \right).$$

Conditioning first upon $\mathscr{G}_{[NT]}^N$, then upon $\mathscr{G}_{[NT]-1}^N$, etc., and using repeatedly the last computations, we deduce that, provided $2\varepsilon_K^2 C_{N,K} \leq N$,

$$\mathbb{P}\left(\sup_{i<NT} |\tilde{M}_{i+1}^{N,K} - \tilde{M}_i^{N,K}| > \varepsilon \right) \leq 1 - \left(1 - \frac{C_{N,K}}{\varepsilon_K^2 N} \right)^{[NT]}$$

$$\leq 1 - \exp(-(2\log 2)\varepsilon_K^{-2} C_{N,K} T).$$

It remains to show that for each $K > 0$, $C_{N,K} \to 0$, as $N \to \infty$. This follows readily from the fact that our assumption (3.2) allows us to deduce from Lindeberg's theorem that if $X_N := \frac{1}{\sqrt{N}} \sum_{j=1}^N \bar{\xi}_j^{N,i}$, $X_N \Rightarrow X$ as $N \to \infty$, where X is a centered normal r.v. with variance 2. But we also have that $\mathbb{E}[X_N^2] \to \mathbb{E}[X^2]$, hence the collection of r.v.'s $\{X_N^2, N \geq 1\}$ is uniformly integrable, from which we deduce that $C_{N,K} \to 0$, as $N \to \infty$. $\square$

3.2 The Individuals with an Infinite Line of Descent

Consider again the collection indexed by N of BGW processes $\{X_n^{N,x}\}$ introduced at the beginning of the previous section, but this time with $\gamma_N = \gamma > 0$, $\sigma_N = \sigma$, for all $N \geq 1$. For each $t \geq 0$, let Y_t^N denote the individuals in the population $X_{[Nt]}^{N,x}$ with an infinite line of descent. Let us describe the law of Y_0^N. Each of the $[Nx]$ individuals living at time $t = 0$ has the probability $1 - q_N$ of having an infinite line of descent, if q_N is the probability of extinction for a population with a unique ancestor at the generation 0. It then follows from the branching property that the law of Y_0^N is the binomial law $B([Nx], 1 - q_N)$. It remains to evaluate q_N, the unique solution in the interval $(0,1)$ of the equation $f_N(x) = x$. Note that

$$f_N''(1) = \mathbb{E}[\xi_N(\xi_N - 1)] = \sigma^2 + \frac{\gamma}{N} + \left(\frac{\gamma}{N}\right)^2.$$

We deduce from a Taylor expansion of f near $x = 1$ that

$$1 - q_N = \frac{2\gamma}{\sigma^2 N} + \circ\left(\frac{1}{N}\right).$$

Consequently

Proposition 5. *In the above model, the number Y_0^N of individuals at time 0 with an infinite line of descent converges in law, as $N \to \infty$, towards $Poi(2x\gamma/\sigma^2)$.*

3.3 Convergence of Continuous Time Branching Processes

Consider a continuous time $\mathbb{Z}_+$-valued branching process $X_t^{N,x}$, with initial condition $X_0^{N,x} = [Nx]$ and reproduction measure μ_N such that $\mu_N(\mathbb{N}) = N$,

$$N^{-1} \sum_{k \geq 0} k\mu_N(k) = 1 + \frac{\gamma_N}{N}, \quad Var(N^{-1}\mu_N) = \sigma_N^2,$$

with $\gamma_N \to \gamma$ and $\sigma_N \to \sigma$, as $N \to \infty$. We define $Z_t^{N,x} = N^{-1}X_t^{N,x}$.
    $\{P(t), t \geq 0\}$ being a standard Poisson process, let

$$Q_t^N = P\left(N^2 \int_0^t Z_s^{N,x} ds\right).$$

It is fair to decide that Q_t^N is the number of birth events which have happened between time 0 and time t, since $N^2 Z_t^{N,x} = N X_t^{N,x}$ is the rate at which birth events occur. Now

$$Z_t^{N,x} = \frac{[Nx]}{N} + N^{-1} \sum_{n=1}^{Q_t^N} (\xi_n^N - 1),$$

where ξ_n^N denotes the number of offsprings at the n-th birth event. Those constitute an i.i.d. sequence with the common law $N^{-1}\mu_N$, which is globally independent of the Poisson process $P(t)$. We have

$$\begin{aligned}
Z_t^{N,x} &= \frac{[Nx]}{N} + \frac{\gamma_N}{N^2} Q_t^N + N^{-1} \sum_{n=1}^{Q_t^N} (\xi_n^N - \mathbb{E}\xi_n^N) \\
&= \frac{[Nx]}{N} + \gamma_N \int_0^t Z_s^{N,x} ds + \gamma_N \left[N^{-2} Q_t^N - \int_0^t Z_s^{N,x} ds \right] \\
&\quad + N^{-1} \sum_{n=1}^{Q_t^N} (\xi_n^N - \mathbb{E}\xi_n^N) \\
&= \frac{[Nx]}{N} + \gamma_N \int_0^t Z_s^{N,x} ds + \varepsilon_N(t) + M_t^N,
\end{aligned} \tag{3.6}$$

where $\varepsilon_N(t) \to 0$ and M_t^N are martingales, and their quadratic variations satisfy

$$[M^N]_t = N^{-2} \sum_{n=1}^{Q_t^N} (\xi_n^N - \mathbb{E}\xi_n^N)^2,$$

$$\langle M^N \rangle_t = \sigma_N^2 \int_0^t Z_s^{N,x} ds, \tag{3.7}$$

hence

$$\mathbb{E}\langle M^N \rangle_t = \sigma_N^2 \int_0^t \mathbb{E}[Z_s^{N,x}] ds, \tag{3.8}$$

while

$$\mathbb{E}\langle \varepsilon_N \rangle_t = N^{-2} \gamma_N^2 \int_0^t \mathbb{E}[Z_s^{N,x}] ds. \tag{3.9}$$

From (3.6),

$$\mathbb{E}[Z_t^{N,x}] \le x e^{\gamma_N t}.$$

And from this, (3.6), (3.8), and (3.9), we deduce that for all $T > 0$,

$$\sup_{N \ge 1} \mathbb{E}\left(\sup_{0 \le t \le T} Z_t^{N,x} \right) < \infty.$$

It is plain that $\varepsilon_N(t) \to 0$ in probability locally uniformly in t. It follows from these last statements, (3.6) and (3.7), Propositions 37 and 38 that the sequence $\{Z_t^{N,x}, t \ge 0\}$ is tight in $D([0, +\infty))$, and moreover that any limit of a converging subsequence is a solution of the SDE (3.4). In other words we have the

Proposition 6. $Z^{N,x} \Rightarrow Z^x$ as $N \to \infty$ for the topology of locally uniform convergence, where Z^x is the unique solution of the following Feller SDE

$$Z_t^x = x + \gamma \int_0^t Z_r^x dr + \sigma \int_0^t \sqrt{Z_r^x} dB_r, \ t \geq 0.$$

3.4 Convergence of Continuous Time Binary Branching Processes

We now restrict ourselves to continuous time binary branching processes. We refer to subsection 2.2.2, and consider for each $N \geq 1$ a continuous time $\mathbb{Z}_+$-valued Markov birth and death process $X_t^{N,x}$ with birth rate $b_N = \sigma^2 N/2 + \alpha$ and death rate $d_N = \sigma^2 N/2 + \beta$, where $\alpha, \beta \geq 0$, and initial condition $X_0^{N,x} = [Nx]$.

We define $Z_t^{N,x} = N^{-1} X_t^{N,x}$. It is not hard to see that there exist two mutually independent standard (i.e., rate 1) Poisson processes $P_b(t)$ and $P_d(t)$, such that

$$Z_t^{N,x} = \frac{[Nx]}{N} + N^{-1} P_b \left(\left(\frac{\sigma^2}{2} N + \alpha \right) \int_0^t N Z_s^{N,x} ds \right)$$

$$- N^{-1} P_d \left(\left(\frac{\sigma^2}{2} N + \beta \right) \int_0^t N Z_s^{N,x} ds \right).$$

Define the two martingales $M_b(t) = P_b(t) - t$ and $M_d(t) = P_d(t) - t$. We have

$$Z_t^{N,x} = \frac{[Nx]}{N} + (\alpha - \beta) \int_0^t Z_s^{N,x} ds + M^N(t), \text{ where}$$

$$M^N(t) = N^{-1} \left[M_b \left(\left(\frac{\sigma^2}{2} N + \alpha \right) \int_0^t N Z_s^{N,x} ds \right) - M_d \left(\left(\frac{\sigma^2}{2} N + \beta \right) \int_0^t N Z_s^{N,x} ds \right) \right].$$

Consequently its quadratic variation is given as

$$[M^N]_t = N^{-2} \left[P_b \left(\left(\frac{\sigma^2}{2} N + \alpha \right) \int_0^t N Z_s^{N,x} ds \right) + P_d \left(\left(\frac{\sigma^2}{2} N + \beta \right) \int_0^t N Z_s^{N,x} ds \right) \right],$$

$$\langle M^N \rangle_t = \left(\sigma^2 + \frac{\alpha + \beta}{N} \right) \int_0^t Z_s^{N,x} ds.$$

It is plain that $\mathbb{E}(Z_t^{N,x}) \leq x \exp((\alpha - \beta)t)$, and moreover that for any $T > 0$, $\sup_{N \geq 1} \mathbb{E} \left(\sup_{0 \leq t \leq T} Z_t^{N,x} \right) < \infty$. We deduce from the above arguments

Proposition 7. $Z^{N,x} \Rightarrow Z^x$ as $N \to \infty$ for the topology of locally uniform convergence, where Z^x is the unique solution of the following Feller SDE

$$Z_t^x = x + \gamma \int_0^t Z_r^x dr + \sigma \int_0^t \sqrt{Z_r^x} dB_r, \ t \geq 0,$$

where $\gamma = \alpha - \beta$.

3.5 Convergence to an ODE

It can be noted that the conditions for convergence towards a Feller diffusion are rather rigid. In the last case which we considered, we need order (N) intensities for both Poisson processes, with a difference in the intensities which is allowed to be of order 1 only. Consider again the case of a continuous time binary branching process as in the previous section, but this time we assume that the birth rate is constant equal to α, and the death rate is constant equal to β. Assume again that $X_0^{N,x} = [Nx]$, and define as above $Z_t^{N,x} = N^{-1} X_t^{N,x}$. Then

$$Z_t^{N,x} = \frac{[Nx]}{N} + N^{-1} P_b \left(\alpha N \int_0^t Z_s^{N,x} ds \right) - N^{-1} P_d \left(\beta N \int_0^t Z_s^{N,x} ds \right).$$

With again $M_b(t) = P_b(t) - t$ and $M_d(t) = P_d(t) - t$,

$$Z_t^{N,x} = \frac{[Nx]}{N} + (\alpha - \beta) \int_0^t Z_s^{N,x} ds + M^N(t), \quad \text{where}$$

$$M^N(t) = N^{-1} \left[M_b \left(\alpha N \int_0^t Z_s^{N,x} ds \right) - M_d \left(\beta N \int_0^t Z_s^{N,x} ds \right) \right].$$

Now

$$[M^N]_t = N^{-2} \left[P_b \left(\alpha N \int_0^t Z_s^{N,x} ds \right) + P_d \left(\beta N \int_0^t Z_s^{N,x} ds \right) \right],$$

$$\langle M^N \rangle_t = \frac{\alpha + \beta}{N} \int_0^t Z_s^{N,x} ds.$$

We again have that $\mathbb{E}(Z_t^{N,x}) \le x \exp((\alpha - \beta)t)$, and moreover that for any $T > 0$, $\sup_{N \ge 1} \mathbb{E} \left(\sup_{0 \le t \le T} Z_t^{N,x} \right) < \infty$. For any $t > 0$, $\mathbb{E}[(M_t^N)^2] \to 0$ as $N \to \infty$. Hence we have the following law of large numbers:

Proposition 8. *As $N \to \infty$, $Z_t^{N,x}$ converges in probability, locally uniformly in t, towards the solution of the ODE*

$$\frac{dZ_t^x}{dt} = (\alpha - \beta) Z_t^x, \quad Z_0^x = x.$$

Chapter 4
Continuous State Branching Process (CSBP)

In the case of a $\mathbb{Z}_+$-valued process, the branching property says that the progeny of the various individuals alive at time t are i.i.d. In the case of a CSBP, an $\mathbb{R}_+$-valued branching process, the branching property says that for all $x, y > 0$, $\{Z_t^{x+y} - Z_t^x, t \geq 0\}$ is independent of $\{Z_t^{x'}, t \geq 0, 0 < x' \leq x\}$, and has the same law as $\{Z_t^y, t \geq 0\}$.

In this chapter, we want to study CSBPs $(Z_t^x, t \geq 0)_{x \geq 0}$ which are the limits of $\mathbb{Z}_+$-valued branching processes considered in the previous chapter, and which are such that for each fixed $x > 0$, there exists a Brownian motion $\{B_t, t \geq 0\}$ such that

$$Z_t^x = x + \gamma \int_0^t Z_s^x ds + \sigma \int_0^t \sqrt{Z_s^x} dB_s.$$

However, the Brownian motion B in this SDE is not the same for all $x > 0$. The nice way of writing an SDE which is valid for all x has been invented recently by Dawson and Li [15] and is as follows.

Given a unique space-time white noise $W(ds, du)$, Dawson and Li [15] show that the SDE

$$Z_t^x = x + \gamma \int_0^t Z_s^x ds + \sigma \int_0^t \int_0^{Z_s^x} W(ds, du)$$

has unique solution for each $x > 0$, and the so-defined random field $\{Z_t^x, t \geq 0, x \geq 0\}$ is a CSBP. We will explain this result in detail in the next section.

Remark 2. This last equation describes all CSBPs with continuous paths. However, a CSBP may have discontinuities. All CSBPs are solution of an SDE of the type

$$Z_t^x = x + \gamma \int_0^t Z_s^x ds + \sigma \int_0^t \int_0^{Z_s^x} W(ds, du) + \int_0^t \int_0^{Z_s^{x-}} \int_0^1 z \bar{M}(ds, dz, du)$$
$$+ \int_0^t \int_0^{Z_s^{x-}} \int_1^\infty z M(ds, dz, du),$$

© Springer International Publishing Switzerland 2016
É. Pardoux, *Probabilistic Models of Population Evolution*, Mathematical
Biosciences Institute Lecture Series 1.6, DOI 10.1007/978-3-319-30328-4_4

where $M(ds,dz,du)$ is a Poisson Point Measure on $(\mathbb{R}_+)^3$ with mean measure $ds \times \pi(dz) \times du$ and $\bar{M}(ds,dz,du) = M(ds,dz,du) - ds \times \pi(dz) \times du$ is the compensated PPM. Here the measure π is assumed to satisfy $\pi(0) = 0$ and $\int_0^\infty (1 \wedge z^2)\pi(dz) < \infty$. We will not discuss those discontinuous CSBPs in these notes. We refer the interested reader to [15] for discussion of the above SDE and its connection with CSBPs. Note also a connection with Lévy processes which is described by the Lamperti transform (see [25]), and hence with the Lévy–Khintchine formula, see Theorem 20 below. Let us state the Lamperti transform, since in the above framework it is intuitively rather clear. Define $A_t^x = \int_0^t Z_s^x ds$, $\tau_t^x = \inf\{s > 0,\ A_s^x > t\}$ and $X_t^x = Z_{\tau_t^x}^x$. Then X_t^x is a Lévy process of the form

$$X_t^x = x + \gamma t + \sigma B_t + \int_0^t \int_0^1 z\bar{N}(ds,dz) + \int_0^t \int_1^\infty zN(ds,dz)$$

stopped at the first time it hits 0, where N is a PPM on $\mathbb{R}_+^2$ with mean measure $ds \times \pi(dz)$. We refer the reader to [13] and [25] for a proof of that result.

4.1 Space-Time White Noise, Dawson–Li SDE, and the Branching Property

Let us first define the space-time white noise on $\mathbb{R}_+^2$. We consider a generalized centered Gaussian random field $\{W(h),\ h \in L^2(\mathbb{R}_+^2)\}$ with covariance $\mathbb{E}[W(h)W(k)] = \langle h,k \rangle$, where $\langle \cdot,\cdot \rangle$ denotes the scalar product in $L^2(\mathbb{R}_+^2)$. In particular, if h and k have disjoint supports, then $W(h)$ and $W(k)$ are independent. An alternative way of writing $W(h)$ is

$$W(h) = \int_0^\infty \int_0^\infty h(t,u)W(dt,du).$$

This can be termed a Wiener integral, i.e., the integral of a deterministic square integrable function. We now want to define an Itô integral. Here the two variables t and u will play very asymmetric roles. Indeed, we will integrate random functions which are adapted in the t direction only. More precisely, let for all $t \geq 0$ $\mathscr{F}_t = \sigma\{W(h),\ \mathrm{supp}(h) \subset [0,t] \times \mathbb{R}_+\} \vee \mathscr{N}$, where $\mathscr{N}$ stands for the σ-algebra of $\mathbb{P}$-null sets, and let $\mathscr{P}$ denote the associated σ-algebra of progressively measurable subsets of $\mathbb{R}_+ \times \Omega$, i.e., $\mathscr{P}$ is generated by the sets of the form $(s,t] \times A$, where $0 \leq s < t$ and $A \in \mathscr{F}_s$.

We now consider random fields of the form

$$\psi : \mathbb{R}_+ \times \Omega \times \mathbb{R}_+,$$

which are assumed to be $\mathscr{P} \otimes \mathscr{B}_+$, where $\mathscr{B}_+$ denotes the σ-algebra of Borel subsets of $\mathbb{R}_+$, and satisfy the assumption that for all $t > 0$,

$$\mathbb{E}\int_0^t \int_0^\infty \psi^2(s,u)dsdu < \infty, \text{ or at least } \int_0^t \int_0^\infty \psi^2(s,u)dsdu < \infty \text{ a.s.}$$

For such ψ's, the stochastic integral

$$\int_0^t \int_0^\infty \psi(s,u)W(ds,du)$$

can be constructed as the limit in probability of the approximating sequence

$$n^2 \sum_{i=1}^{[nt]-1} \sum_{j=1}^\infty \langle \psi, \mathbf{1}_{A_{i-1,j}^n} \rangle W(A_{i,j}^n),$$

where we have used the abuse of notation $W(A) = W(\mathbf{1}_A)$, and

$$A_{i,j}^n = \left[\frac{i}{n}, \frac{i+1}{n} \right] \times \left[\frac{j}{n}, \frac{j+1}{n} \right].$$

The resulting process $\{ \int_0^t \int_0^\infty \psi(s,u)W(ds,du), \, t \geq 0 \}$ is a continuous local martingale, which satisfies

$$\mathbb{E}\left[\left(\int_0^t \int_0^\infty \psi(s,u)W(ds,du) \right)^2 \right] \leq \mathbb{E} \int_0^t \int_0^\infty \psi^2(s,u)dsdu,$$

with equality whenever the right-hand side is finite, in which case the above local martingale is a square integrable martingale. Note also that

$$\langle M \rangle_t = \int_0^t \int_0^\infty \psi^2(s,u)duds.$$

We refer to [43] for a detailed construction of that stochastic integral.

We now turn to the Dawson–Li SDE

$$Z_t^x = x + \gamma \int_0^t Z_s^x ds + \sigma \int_0^t \int_0^\infty \mathbf{1}_{u \leq Z_s^x} W(ds,du). \tag{4.1}$$

We now prove existence and uniqueness of a strong solution of that equation.

Theorem 1. *Given the space-time white noise W, equation* (4.1) *has a unique strong solution.*

PROOF: STEP 1 We first prove pathwise uniqueness. Consider the decreasing sequence of positive numbers $a_k = \exp(-k(k+1)/2)$, $k \geq 0$, so that $a_0 = 1$, $\int_{a_k}^{a_{k-1}} x^{-1}dx = k$ and $a_k \downarrow 0$ as $k \to \infty$. Now for each $k \geq 1$, let ψ_k be a continuous function defined on $\mathbb{R}_+$ such that

$$\text{supp}(\psi_k) \subset [a_k, a_{k-1}], \; 0 \leq \psi_k(x) \leq \frac{2}{kx}, \; \int_{a_k}^{a_{k-1}} \psi_k(x)dx = 1,$$

and let

$$\phi_k(x) = \int_0^{|x|} dy \int_0^y \psi_k(z)dz.$$

We note that ϕ_k is of class C^2, $0 \le \phi_k'(x) \le 1$ for $x > 0$, $-1 \le \phi_k'(x) \le 0$ for $x < 0$, and $0 \le \phi_k(x) \uparrow |x|$, as $k \to \infty$. Let Z_t^1 and Z_t^2 denote two solutions of the SDE (4.1). It follows from Itô's formula that for any $k \ge 1$,

$$
\mathbb{E}\phi_k(Z_t^1 - Z_t^2) = \gamma \mathbb{E} \int_0^t \phi_k'(Z_s^1 - Z_s^2)(Z_s^1 - Z_s^2)ds
$$

$$
+ \frac{1}{2}\mathbb{E} \int_0^t \phi_k''(Z_s^1 - Z_s^2)|Z_s^1 - Z_s^2|ds
$$

$$
\le |\gamma|\mathbb{E} \int_0^t |Z_s^1 - Z_s^2|ds + \frac{t}{k},
$$

since $0 \le \phi_k'(x)x \le |x|$ and $0 \le \phi_k''(x)|x| \le \frac{2}{k}$. Letting $k \to \infty$, we deduce from the monotone convergence theorem that

$$
\mathbb{E}|Z_s^1 - Z_s^2| \le |\gamma|\mathbb{E} \int_0^t |Z_s^1 - Z_s^2|ds,
$$

from which $\mathbb{E}|Z_t^1 - Z_t^2| = 0$ for all $t \ge 0$ follows from Gronwall's Lemma.

STEP 2 Existence of a weak solution follows from the results of chapter 3 and the comment at the end of this subsection. Strong existence now follows from pathwise uniqueness, due to an argument of Yamada–Watanabe, see, e.g., section 3.9 in [35]. □

Remark 3. As already noticed, there exists a Brownian motion B_t such that (4.1) can be rewritten as

$$
Z_t^x = x + \gamma \int_0^t Z_s^x ds + \sigma \int_0^t \sqrt{Z_s^x}dB_s. \tag{4.2}
$$

The last proof is an adaptation of Yamada–Watanabe's proof of strong uniqueness for this last equation, see [44]. Note that while the Itô formula for $\phi(Z_t^x)$ gives exactly the same "additional Itô term" in both formulations of the SDE, this is no longer true for $\phi(Z_t^1 - Z_t^2)$. The bracket of the martingale part of the solution is the same in both formulations, but not the bracket of the difference between two solutions. However the proof is essentially the same in both cases.

As a matter of fact, the same argument as in the proof of Theorem 1 allows us to prove

Corollary 1. *For any $x > 0$, equation (4.2) has a unique strong solution.*

We now have the comparison theorem

Theorem 2. *Let Z_t^x and Z_t^y be two solutions of (4.1), with the initial conditions $Z_0^x = x$, $Z_0^y = y$, and assume that $x < y$. Then $\mathbb{P}(Z_t^x \le Z_t^y$ for all $t > 0) = 1$.*

PROOF: We introduce the same sequence $(\psi_k)_{k \ge 1}$ as in the proof of Theorem 1, and define this time

$$
\phi_k(x) = \int_0^x dy \int_0^y \psi_k(z)dz.
$$

Again ϕ_k is of class C^2, $\phi'_k(x) = 0$ for $x < 0$, $0 \leq \phi'(x) \leq 1$ if $x > 0$, and $0 \leq \phi_k(x) \uparrow x$, as $k \to \infty$. We deduce from Itô's formula

$$\mathbb{E}\phi_k(Z^x_t - Z^y_t) = \gamma\mathbb{E}\int_0^t \phi'_k(Z^x_s - Z^y_s)(Z^x_s - Z^y_s)ds$$

$$+ \frac{1}{2}\mathbb{E}\int_0^t \phi''_k(Z^x_s - Z^y_s)|Z^x_s - Z^y_s|ds$$

$$\leq |\gamma|\mathbb{E}\int_0^t (Z^x_s - Z^y_s)_+ ds + \frac{t}{k}.$$

Letting $k \to \infty$, we deduce from the monotone convergence theorem that

$$\mathbb{E}[(Z^x_s - Z^y_s)_+] \leq |\gamma|\mathbb{E}\int_0^t (Z^x_s - Z^y_s)_+ ds,$$

from which $\mathbb{E}[(Z^x_t - Z^y_t)_+] = 0$ for all $t \geq 0$ follows from Gronwall's Lemma. The result follows readily. □

As a consequence of the above results, for any $n \geq 1$, any $0 = x_0 < x_1 < x_2 < \cdots < x_n$, the processes $\{Z^{x_k}_t - Z^{x_{k-1}}_t, t \geq 0; 1 \leq k \leq n\}$ are $\mathbb{R}_+$-valued. Moreover, since they are functionals of the white noise W on disjoint subsets of $\mathbb{R}_+ \times \mathbb{R}_+$, they are independent, and it is easy to check that for each $1 \leq k \leq n$, $\{Z^{x_k}_t - Z^{x_{k-1}}_t, t \geq 0\}$ has the same law as $\{Z^{x_k - x_{k-1}}_t, t \geq 0\}$. In other words, the increments of the mapping $x \to Z^x$ are independent and stationary. Consequently the solution of (4.1) is a continuous space branching process.

Since both the approximations and the limit have stationary and independent increments, the three Propositions 4, 6, and 7 can be strengthened as convergence results of $\{Z^{N,x}_t, t \geq 0, x > 0\}$ towards the solution of (4.1). The appropriate topology in the space of two-parameter processes for which these convergences hold will be specified below in the statement of Theorem 13.

4.2 Laplace Functional of a CSBP

The branching property entails that for all t, $\lambda > 0$, there exists a positive real number $u(t,\lambda)$ such that

$$\mathbb{E}[\exp(-\lambda Z^x_t)] = \exp[-xu(t,\lambda)]. \tag{4.3}$$

From the Markov property of the process $t \to Z^x_t$, we deduce readily the semigroup identity

$$u(t+s,\lambda) = u(t,u(s,\lambda)).$$

We seek a formula for $u(t,\lambda)$. Let us first get by a formal argument an ODE satisfied by $u(\cdot,\lambda)$. For $t > 0$ small, we have that

$$Z^x_t \simeq x + \gamma x t + \sigma\sqrt{x}B_t,$$

hence

$$\mathbb{E}\left(e^{-\lambda Z_t^x}\right) \simeq \exp\left(-\lambda x[1 + \gamma t - \sigma^2 \lambda t/2]\right),$$

and

$$\frac{u(t,\lambda) - \lambda}{t} \simeq \gamma \lambda - \frac{\sigma^2}{2}\lambda^2.$$

Assuming that $t \to u(t,\lambda)$ is differentiable, we deduce that

$$\frac{\partial u}{\partial t}(0,\lambda) = \gamma \lambda - \frac{\sigma^2}{2}\lambda^2.$$

This, combined with the semigroup identity, entails that

$$\frac{\partial u}{\partial t}(t,\lambda) = -\Psi(u(t,\lambda)), \quad u(0,\lambda) = \lambda, \tag{4.4}$$

where $\Psi(r) := \sigma^2 r^2/2 - \gamma r$ is called the *branching mechanism* of the continuous state branching process Z. It is easy to solve that ODE explicitly, and we now prove rigorously that u is indeed the solution of (4.4), without having to go through the trouble of justifying the above argument.

Lemma 5. *The function $(t,\lambda) \to u(t,\lambda)$ defined by (4.3) is given by the formula*

$$u(t,\lambda) = \frac{\lambda \gamma e^{\gamma t}}{\gamma + \frac{\lambda \sigma^2}{2} \times (e^{\gamma t} - 1)} \tag{4.5}$$

and it is the unique solution of (4.4).

PROOF: It suffices to show that $\{M_s^x, \ 0 \le s \le t\}$ defined by

$$M_s^x = \exp\left(-\frac{\gamma e^{\gamma(t-s)}}{\sigma^2(e^{\gamma(t-s)} - 1)/2 + \gamma/\lambda} Z_s^x\right)$$

is a martingale, which follows from Itô's formula.   □

Remark 4. In the critical case (i.e., the case $\gamma = 0$),

$$u(t,\lambda) = \frac{\lambda}{1 + \sigma^2 \lambda t/2},$$

which is the limit as $\gamma \to 0$ of (4.5). This particular formula can also be established by checking that in the case $\gamma = 0$,

$$N_s^x = \exp\left(-\frac{\lambda}{1 + \sigma^2 \lambda(t-s)/2} Z_s^x\right)$$

is a martingale.

For each fixed $t > 0$, $x \to Z_t^x$ has independent and homogeneous increments with values in $\mathbb{R}_+$. We shall consider its right-continuous modification, which then is a subordinator. Its Laplace exponent is the function $\lambda \to u(t, \lambda)$, which can be rewritten (like for any subordinator, see section A.6 below) as

$$u(t, \lambda) = d(t)\lambda + \int_0^\infty (1 - e^{-\lambda r})\Lambda(t, dr),$$

where $d(t) \geq 0$ and for each $t > 0$, the measure $\Lambda(t, \cdot)$ satisfies $\int_0^\infty (r \wedge 1)\Lambda(t, dr) < \infty$. Comparing with (4.5), we first deduce $d(t) \equiv 0$ from the fact that $u(t, \infty) < \infty$, and moreover, with the notation $\gamma_t = \frac{2\gamma}{\sigma^2(1 - e^{-\gamma t})}$,

$$\Lambda(t, dr) = p(t)\exp(-q(t)r)dr, \text{ where } p(t) = \gamma_t^2 e^{-\gamma t}, \ q(t) = \gamma_t e^{-\gamma t}. \tag{4.6}$$

We have defined the two-parameter process $\{Z_t^x; \ x \geq 0, t \geq 0\}$. Z_t^x is the population at time t made of descendants of the initial population of size x at time 0. We may want to introduce three parameters, if we want to discuss the descendants at time t of a population of a given size at time s. The first point, which is technical but in fact rather standard, is that we can construct the collection of those random variables jointly for all $0 \leq s < t, x \geq 0$, so that all the properties we may reasonably wish for them are satisfied. More precisely, following [7], we have the

Lemma 6. *On some probability space, there exists a three parameter process*

$$\{Z_{s,t}^x, \ 0 \leq s \leq t, \ x \geq 0\},$$

such that

1. *For every* $0 \leq s \leq t$, $Z_{s,t} = \{Z_{s,t}^x, \ x \geq 0\}$ *is a subordinator with Laplace exponent* $u(t - s, \cdot)$.
2. *For every* $n \geq 2$, $0 \leq t_1 < t_2 < \cdots < t_n$, *the subordinators* $Z_{t_1, t_2}, \ldots, Z_{t_{n-1}, t_n}$ *are mutually independent, and*

$$Z_{t_1, t_n}^x = Z_{t_{n-1}, t_n} \circ \cdots \circ Z_{t_1, t_2}^x, \ \forall x \geq 0, \ a.s.$$

3. *The processes* $\{Z_{0,t}^x, \ t \geq 0, x \geq 0\}$ *and* $\{Z_t^x, \ t \geq 0, x \geq 0\}$ *have the same finite dimensional distributions.*

Now consider $\{Z_{s,t}^x, \ x \geq 0\}$ for fixed $0 \leq s \leq t$. It is a subordinator with Laplace exponent (the functions p and q are given in (4.6))

$$u(t - s, \lambda) = p(t - s)\int_0^\infty (1 - e^{-\lambda r})e^{-q(t-s)r}dr.$$

We shall give a probabilistic description of the process $\{Z_{s,t}^x, \ x \geq 0\}$ below in section 4.4. From now on, we write Z_t^x for $Z_{0,t}^x$.

Let us first study the large time behavior of the process Z_t^x. This CSBP is said to be subcritical if $\gamma < 0$ critical if $\gamma = 0$ and supercritical if $\gamma > 0$. Consider the extinction event

$$E = \{\exists t > 0, \text{ s. t. } Z_t^x = 0\}.$$

Proposition 9. *If* $\gamma \leq 0$, $\mathbb{P}_x(E) = 1$ *a.s. for all* $x > 0$. *If* $\gamma > 0$, $\mathbb{P}_x(E) = \exp(-2x\gamma/\sigma^2)$ *and on* E^c, $Z_t^x \to +\infty$ *a.s.*

Remark 5. Recall the result of Proposition 5, and the fact that a $\text{Poi}(2x\gamma/\sigma^2)$ r.v. is zero with probability $\exp(-2x\gamma/\sigma^2)$.

PROOF: If $\gamma \leq 0$, $\{Z_t^x, t \geq 0\}$ is a positive supermartingale. Hence it converges a.s. as $t \to \infty$. The limit r. v. Z_∞^x takes values in the set of fixed points of the SDE (3.4), which is $\{0, +\infty\}$. But from Fatou and the supermartingale property,

$$\mathbb{E}(\lim_{t \to \infty} Z_t^x) \leq \lim_{t \to \infty} \mathbb{E}(Z_t^x) \leq x.$$

Hence $\mathbb{P}(Z_\infty^x = +\infty) = 0$, and $Z_t^x \to 0$ a.s. as $t \to \infty$.

If now $\gamma > 0$, it follows from Itô's formula that $M_t = \exp(-\frac{2\gamma}{\sigma^2} Z_t^x)$ is a martingale with values in $[0,1]$, which converges a.s. as $t \to \infty$. Consequently $Z_t^x = -\sigma^2 \log(M_t)/2\gamma$ converges a.s., and as above its limit belongs to the set $\{0, +\infty\}$. Moreover

$$\mathbb{P}(E) = \lim_{t \to \infty} \mathbb{P}(Z_t^x = 0)$$

$$= \lim_{t \to \infty} \mathbb{E}[\exp\{-xu(t, \infty)\}]$$

$$= \lim_{t \to \infty} \exp\left\{-x\frac{2\gamma e^{\gamma t}}{\sigma^2(e^{\gamma t} - 1)}\right\}$$

$$= \exp\{-2x\gamma/\sigma^2\}.$$

It remains to prove that

$$\mathbb{P}(E^c \cap \{Z_t \to 0\}) = 0. \tag{4.7}$$

Define the stopping times

$$\tau_1 = \inf\{t > 0, Z_t^x \leq 1\}, \text{ and for } n \geq 2,$$
$$\tau_n = \inf\{t > \tau_{n-1} + 1, Z_t^x \leq 1\}.$$

On the set $\{Z_t^x \to 0\}$, as $t \to \infty$, $\tau_n < \infty$, $\forall n$. Define for $n \geq 1$

$$A_n = \{\tau_{n+1} < \infty, Z_{\tau_{n+1}}^x > 0\}.$$

For all $N > 0$,

$$\mathbb{P}(E^c \cap \{Z_t \to 0\}) \leq \mathbb{P}(\cap_{n=1}^N A_n)$$

$$\leq \mathbb{E}\left(\prod_{n=1}^N \mathbb{P}(A_n|\mathscr{F}_{\tau_n})\right)$$

$$\leq (\mathbb{P}(Z_1(1) > 0))^N$$

$$\to 0, \text{ as } N \to \infty,$$

where we have used the strong Markov property, and the fact that

$$\mathbb{P}(A_n|Z_{\tau_n}) \leq \mathbb{P}(Z_1(1) > 0).$$

□

4.3 The Individuals Whose Progeny Survives During tN Generations

In this section, we assume that $\gamma > 0$ (Z_t^x is supercritical). Proposition 5 indicates that if we consider only the *prolific individuals*, i.e., those with an infinite line of descent, in the limit $N \to \infty$, we should not divide their number by N, also $X_{[Nt]}^{N,x} \to +\infty$, as $N \to \infty$, for all $t \geq 0$. If we now consider those individuals whose progeny is still alive at time tN (i.e., those whose progeny contributes to the population at time $t > 0$ in the limit as $N \to \infty$), then again we should not divide by N. We shall explain at the start of the next section how this follows from (we use again the notation $\gamma_t = 2\gamma[\sigma^2(1 - e^{-\gamma t})]^{-1}$)

Theorem 3. *Under the assumptions from the beginning of section 3.1, with the understanding that under $\mathbb{P}_1$, $X_0^N = 1$,*

1. for N large,

$$\mathbb{P}_1(X_{[Nt]}^N > 0) = \frac{\gamma_t}{N} + \circ\left(\frac{1}{N}\right),$$

and
2. as $N \to \infty$,

$$\mathbb{E}_1\left(\exp[-\lambda X_{[Nt]}^N/N]\Big|X_{[Nt]}^N > 0\right) \to \frac{\gamma_t e^{-\gamma t}}{\lambda + \gamma_t e^{-\gamma t}}.$$

PROOF OF 1 : It follows from the branching property that

$$\mathbb{P}_1(X_{[Nt]}^N > 0) = 1 - \mathbb{P}_1(X_{[Nt]}^N = 0)$$

$$= 1 - \mathbb{P}_N(X_{[Nt]}^N = 0)^{1/N}$$

$$= 1 - \mathbb{P}_1(Z_t^N = 0)^{1/N}.$$

But

$$\log\left[\mathbb{P}_1(Z_t^N = 0)^{1/N}\right] = \frac{1}{N}\log\mathbb{P}_1(Z_t^N = 0)$$
$$= \frac{1}{N}\log\mathbb{P}_1(Z_t = 0) + \circ\left(\frac{1}{N}\right).$$

From (4.3) and (4.5), we deduce that

$$\mathbb{P}_1(Z_t = 0) = \lim_{\lambda\to\infty}\exp[-u(t,\lambda)]$$
$$= \exp(-\gamma_t).$$

We then conclude that

$$\mathbb{P}_1(X_{[Nt]}^N > 0) = 1 - \exp\left[-\frac{\gamma_t}{N} + \circ\left(\frac{1}{N}\right)\right]$$
$$= \frac{\gamma_t}{N} + \circ\left(\frac{1}{N}\right).$$

PROOF OF 2 : Using (4.5) again, we obtain

$$\mathbb{E}_1\exp[-\lambda X_{[Nt]}^N/N] = \left(\mathbb{E}_N\exp[-\lambda X_{[Nt]}^N/N]\right)^{1/N}$$
$$\simeq (\mathbb{E}_1\exp[-\lambda Z_t])^{1/N}$$
$$= \exp\left(-\frac{\lambda\gamma_t}{N(\lambda + \gamma_t e^{-\gamma_t})}\right).$$

But

$$\mathbb{E}_1\left(\exp[-\lambda X_{[Nt]}^N/N]|X_{[Nt]}^N > 0\right) = \frac{\mathbb{E}_1\left(\exp[-\lambda X_{[Nt]}^N/N]; X_{[Nt]}^N > 0\right)}{\mathbb{P}_1(X_{[Nt]}^N > 0)}$$
$$= \frac{\mathbb{E}_1\left(\exp[-\lambda X_{[Nt]}^N/N]\right) - 1 + \mathbb{P}_1(X_{[Nt]}^N > 0)}{\mathbb{P}_1(X_{[Nt]}^N > 0)}$$
$$= 1 + \frac{\mathbb{E}_1\left(\exp[-\lambda X_{[Nt]}^N/N]\right) - 1}{\mathbb{P}_1(X_{[Nt]}^N > 0)}$$
$$\to 1 - \frac{\lambda}{\lambda + \gamma_t e^{-\gamma_t}},$$

as $N \to \infty$. The result follows. $\square$

4.4 Consequence for the CSBP

We assume again that $\gamma > 0$. Note that the continuous limit $\{Z_t\}$ has been obtained after a division by N, so that Z_t no longer represents a number of individuals, but a sort of density. The point is that there are constantly infinitely many births and deaths, most individuals having a very short life. If we consider only those individuals at time 0 whose progeny is still alive at some time $t > 0$, that number is finite. We now explain how this follows from the last theorem, and show how it provides a probabilistic description of the subordinator which appeared in section 4.2.

The first part of Theorem 3 tells us that for large N, each of the N individuals from the generation 0 has a progeny at the generation $[Nt]$ with probability $\gamma_t/N + \circ(1/N)$, independently of the others. Hence the number of those individuals tends to the Poisson law with parameter γ_t. The second statement says that those individuals contribute to Z_t a quantity which follows an exponential random variable with parameter $\gamma_t e^{-\gamma t}$. This means that

$$Z_{0,t}^x = \sum_{i=1}^{N_x} Y_i,$$

where N_x, $Y_1, Y_2, \ldots$ are mutually independent, the law of N_x being Poisson with parameter $x\gamma_t$, and the law of each Y_i exponential with parameter $\gamma_t e^{-\gamma t}$.

Taking into account the branching property, we have more precisely that $\{Z_{0,t}^x, \ x \geq 0\}$ is a compound Poisson process, the set of jump locations being a Poisson process with intensity γ_t, the jumps being i.i.d., exponential with parameter $\gamma_t e^{-\gamma t}$. We can recover from this description the formula for the Laplace exponent of Z_t^x. Indeed

$$\mathbb{E}\exp\left(-\lambda\sum_{i=1}^{N_x} Y_i\right) = \sum_{k=0}^{\infty}\left(\mathbb{E}e^{-\lambda Y_1}\right)^k \mathbb{P}(N_x = k)$$

$$= \exp\left(-x\frac{\lambda\gamma_t}{\lambda + \gamma_t e^{-\gamma t}}\right).$$

We can now describe the genealogy of the population whose total mass follows the SDE (3.4).

Suppose that Z ancestors from $t = 0$ contribute, respectively, $Y_1, Y_2, \ldots, Y_Z$ to $Z_{0,t}^x$. This means that the compound Poisson process $\{Z_{0,t}^y, \ y > 0\}$ has Z jumps on the interval $[0,x]$, with respective sizes $Y_1, Y_2, \ldots, Y_Z$. Consider now $Z_{0,t+s}^x = Z_{t,t+s}(Z_{0,t}^x)$. From the Y_1 mass at time t, a finite number Z_1 of individuals, which follows a Poisson law with parameter $Y_1\gamma_s$, has a progeny at time $t + s$, each one contributing an exponential r. v. with parameter $\gamma_s e^{-\gamma s}$ to $Z_{0,t+s}^x$.

For any $y, z \geq 0$, $0 \leq s < t$, we say that the individual z in the population at time t is a descendant of the individual y from the population at time s if y is a jump location of the subordinator $x \to Z_{s,t}^x$, and moreover

$$Z_{s,t}(y^-) < z < Z_{s,t}(y).$$

Note that $\Delta Z_{s,t}(y) = Z_{s,t}(y) - Z_{s,t}(y^-)$ is the contribution to the population at time t of the progeny of the individual y from the population at time s.

4.5 The Prolific Individuals

Once more we assume that $\gamma > 0$. We want to consider again the individuals with an infinite line of descent, but directly in the continuous model. Those could be defined as the individuals such that $\Delta Z_{0,t}^y > 0$, for all $t > 0$. However, it should be clear from Proposition 9 that an a.s. equivalent definition is the following:

Definition 1. The individual y from the population at time s is said to be prolific if $\Delta Z_{s,t}^y \to \infty$, as $t \to \infty$.

For any $s \geq 0$, $x > 0$, let

$$\mathscr{P}_s^x = \{y \in [0, Z_s^x]; \ \Delta Z_{s,t}^y \to \infty, \text{ as } t \to \infty\},$$
$$P_s^x = \text{card}(\mathscr{P}_s^x).$$

We denote as follows the conditional probability given extinction

$$\mathbb{P}_e = \mathbb{P}(\cdot|E) = e^{2x\gamma/\sigma^2}\mathbb{P}(\cdot \cap E).$$

We now have

Proposition 10. *Under $\mathbb{P}_e$, there exists a standard Brownian motion $\{B_t^e, \ t \geq 0\}$ such that $Z_\cdot^x$ solves the SDE*

$$Z_t^x = x - \gamma \int_0^t Z_s^x ds + \sigma \int_0^t \sqrt{Z_s^x}\, dB_s^e.$$

PROOF: Clearly

$$\begin{aligned}
\mathbb{E}_e[f(Z_t^x)] &= \frac{\mathbb{E}[f(Z_t^x); E]}{\mathbb{P}(E)}\\
&= \frac{\mathbb{E}[f(Z_t^x)\mathbb{P}(E|Z_t^x)]}{\mathbb{P}(E)}\\
&= e^{2\gamma x/\sigma^2}\mathbb{E}\left[e^{-2\gamma Z_t^x/\sigma^2} f(Z_t^x)\right].
\end{aligned}$$

This means that the law of Z_t^x under $\mathbb{P}_e$ is absolutely continuous with respect to its law under $\mathbb{P}$, and the Radon–Nikodym derivative is given by

$$e^{-2\gamma Z_t^x/\sigma^2 + 2\gamma x/\sigma^2} = \exp\left\{ -\frac{2\gamma}{\sigma} \int_0^t \sqrt{Z_s^x}\, dB_s - \frac{2\gamma^2}{\sigma^2} \int_0^t Z_s^x\, ds \right\}.$$

The result now follows from Girsanov's theorem, see Proposition 35 in section A.5 below. □

The branching mechanism of Z under $\mathbb{P}_e$ is given by

$$\Psi_e(r) = \frac{\sigma^2}{2} r^2 + \gamma r = \Psi(r + 2\gamma/\sigma^2).$$

Next we identify the conditional law of Z_t^x, given that $P_t^x = n$, for $n \geq 0$.

Proposition 11. *For any Borel measurable $f : \mathbb{R} \to \mathbb{R}_+$,*

$$\mathbb{E}[f(Z_t^x)|P_t^x = n] = \frac{\mathbb{E}_e[f(Z_t^x)(Z_t^x)^n]}{\mathbb{E}_e[(Z_t^x)^n]}.$$

PROOF: In this proof, we use the notation $\gamma' = 2\gamma/\sigma^2$. Recall that the law of P_0^x is Poi($x\gamma'$). Indeed, the law of P_0^x is the asymptotic law of N_x as $t \to \infty$. See also Proposition 5. Clearly from the Markov property of Z^x, the conditional law of P_t^x, given Z_t^x, is Poi($Z_t^x\gamma'$). Consequently for $\lambda > 0$, $0 \leq s \leq 1$,

$$\mathbb{E}\left(\exp[-\lambda Z_t^x] s^{P_t^x}\right) = \mathbb{E}\left(\exp[-\lambda Z_t^x]\exp[-\gamma'(1-s)Z_t^x]\right)$$

$$= \mathbb{E}\left(\exp[-(\lambda + \gamma')Z_t^x]\exp[\gamma' s Z_t^x]\right)$$

$$= \sum_{n=0}^{\infty} \frac{(s\gamma')^n}{n!} \mathbb{E}\left(\exp[-(\lambda + \gamma')Z_t^x](Z_t^x)^n\right).$$

Now define

$$h(t, \lambda, x, n) = \mathbb{E}\left(\exp[-\lambda Z_t^x]|P_t^x = n\right).$$

Note that

$$\mathbb{P}(P_t^x = n) = \mathbb{E}\left[\mathbb{P}(P_t^x = n|Z_t^x)\right]$$

$$= \frac{\gamma'^n}{n!} \mathbb{E}\left(e^{-\gamma' Z_t^x}(Z_t^x)^n\right).$$

Consequently, conditioning first upon the value of P_t^x, and then using the last identity, we deduce that

$$\mathbb{E}\left(\exp[-\lambda Z_t^x] s^{P_t^x}\right) = \sum_{n=0}^{\infty} \frac{(s\gamma')^n}{n!} h(t, \lambda, x, n) \mathbb{E}\left(\exp[-\gamma' Z_t^x](Z_t^x)^n\right).$$

Comparing the two series, and using the fact that, on $\mathscr{F}_t$, $\mathbb{P}_e$ is absolutely continuous with respect to $\mathbb{P}$, with density $e^{x\gamma'}\exp[-\gamma'Z_t^x]$, we deduce that for all $n \geq 0$,

$$h(t,\lambda,x,n) = \frac{\mathbb{E}\left(\exp[-(\lambda+\gamma')Z_t^x](Z_t^x)^n\right)}{\mathbb{E}\left(\exp[-\gamma'Z_t^x](Z_t^x)^n\right)}$$
$$= \frac{\mathbb{E}_e\left(\exp[-\lambda Z_t^x](Z_t^x)^n\right)}{\mathbb{E}_e\left[(Z_t^x)^n\right]}.$$

□

To any probability law v on $\mathbb{R}_+$ with finite mean c, we associate the so-called law of its size-biased picking as the law on $\mathbb{R}_+$ $c^{-1}yv(dy)$. We note that the conditional law of Z_t^x, given that $P_t^x = n+1$ is obtained from the conditional law of Z_t^x, given that $P_t^x = n$ by sized-biased picking.

We now describe the law of $\{P_t^x, t \geq 0\}$, for fixed $x > 0$. Clearly this is a continuous time B–G–W process as considered above in chapter 2. We have the

Theorem 4. *For every $x > 0$, the process $\{P_t^x, t \geq 0\}$ is an $\mathbb{N}$-valued immortal branching process in continuous time, with initial distribution $\mathrm{Poi}(2x\gamma/\sigma^2)$, and reproduction measure μ_P given by*

$$\mu_P(n) = \begin{cases} \gamma, & \text{if } n = 2, \\ 0, & \text{if } n \neq 2. \end{cases}$$

In other words, $\{P_t^x, t \geq 0\}$ is a Yule tree with intensity γ.

Remark 6. If we call Φ_P the Φ-function (with the notations of section 2.2) associated to the measure μ_P, we have in terms of the branching mechanism Ψ of Z

$$\Phi_P(s) = \gamma(s^2 - s) = \frac{\sigma^2}{2\gamma}\Psi\left(\frac{2\gamma}{\sigma^2}(1-s)\right).$$

Note that Ψ_e describes the branching process Z, conditioned upon extinction, while Φ_P describes the immortal part of Z. Φ_P depends upon the values $\Psi(r), 0 \leq r \leq \gamma/2$, while Ψ_e depends upon the values $\Psi(r), \gamma/2 \leq r \leq 1$. The mapping $\Psi \to (\Psi_e, \Phi_P)$ should be compared with the mapping $f \to (\tilde{f}, f^*)$ from Proposition 2.

PROOF: The process P^x inherits its branching property from that of Z^x. The immortal character is obvious. We already know that the conditional law of P_t^x, given Z_t^x, is the Poisson law with parameter $2Z_t^x\gamma/\sigma^2$. Consequently

$$\mathbb{E}\left(s^{P_t^x}\right) = \mathbb{E}\left(\exp[-(1-s)2\gamma Z_t^x/\sigma^2]\right)$$
$$= \exp[-xu(t, 2(1-s)\gamma/\sigma^2)].$$

Moreover, if we call $\psi_t(s)$ the generating function of the continuous time B–G–W process $\{P_t^x, \ t \geq 0\}$, we have that

$$\mathbb{E}\left(s^{P_t^x}\right) = \mathbb{E}\left(\psi_t(s)^{P_0^x}\right)$$
$$= \exp[-2x\gamma(1 - \psi_t(s))/\sigma^2].$$

Comparing those two formulas, we deduce that

$$1 - \psi_t(s) = \frac{\sigma^2}{2\gamma}u(t, 2(1-s)\gamma/\sigma^2).$$

Taking the derivative with respect to the time variable t, we deduce from the differential equations satisfied by $\psi_t(\cdot)$ and by $u(t, \cdot)$ the identity

$$\Phi_P(\psi_t(s)) = \frac{\sigma^2}{2\gamma}\Psi(u(t, 2(1-s)\gamma/\sigma^2)) = \frac{\sigma^2}{2\gamma}\Psi(2\gamma(1-\psi_t(s))/\sigma^2).$$

Consequently

$$\Phi_P(r) = \frac{\sigma^2}{2\gamma}\Psi(2\gamma(1-r)/\sigma^2).$$

The measure μ_P is then recovered easily from Φ_P. $\square$

We next note that the pair (Z_t^x, P_t^x) enjoys the Branching property in the following sense. For every $x > 0$, $n \in \mathbb{N}$, denote by $(Z.(x,n), P.(x,n))$ a version of the process $\{(Z_t^x, P_t^x), \ t \geq 0\}$, conditioned upon $P_0^x = n$. What we mean here by the branching property is the fact that for all $x, x' > 0$, $n, n' \in \mathbb{N}$,

$$(Z.(x+x', n+n'), P.(x+x', n+n'))$$

has the same law as

$$(Z.(x,n), P.(x,n)) + (Z'.(x',n'), P'.(x',n')),$$

where the two processes $(Z.(x,n), P.(x,n))$ and $(Z'.(x',n'), P'.(x',n'))$ are mutually independent.

We now characterize the joint law of $(Z_t(x,n), P_t(x,n))$. Again below $\gamma' = 2\gamma/\sigma^2$.

Proposition 12. *For any $\lambda \geq 0$, $s \in [0,1]$, $t \geq 0$, $x > 0$, $n \in \mathbb{N}$,*

$$\mathbb{E}\left(\exp[-\lambda Z_t(x,n)]s^{P_t(x,n)}\right)$$
$$= \exp[-x(u(t, \lambda + \gamma') - \gamma')]\left(\frac{u(t, \lambda + \gamma') - u(t, \lambda + \gamma'(1-s))}{\gamma'}\right)^n.$$

PROOF: First consider the case $n = 0$. We note that $Z.(x,0)$ is a version of the continuous branching process conditioned upon extinction, i.e., with branching mechanism $\Psi_e(r) = \Psi(r + \gamma')$, while $P_t(x,0) \equiv 0$. Hence

$$\mathbb{E}\left(\exp[-\lambda Z_t(x,0)]s^{P_t(x,0)}\right) = \exp[-x(u(t,\lambda + \gamma') - \gamma')]. \tag{4.8}$$

Going back to the computation at the beginning of the proof of Proposition 11, we have

$$\mathbb{E}\left(\exp[-\lambda Z_t^x]s^{P_t^x}\right) = \mathbb{E}\left(\exp[-(\lambda + \gamma'(1-s))Z_t^x]\right)$$
$$= \exp[-xu(t,\lambda + \gamma'(1-s))].$$

Since the law of P_0^x is Poisson with parameter $x\gamma'$,

$$\mathbb{E}\left(\exp[-\lambda Z_t^x]s^{P_t^x}\right) = \sum_{n=0}^{\infty} e^{-x\gamma'}\frac{(x\gamma')^n}{n!}\mathbb{E}\left(\exp[-\lambda Z_t(x,n)]s^{P_t(x,n)}\right).$$

From the branching property of (Z,P),

$$\mathbb{E}\left(\exp[-\lambda Z_t(x,n)]s^{P_t(x,n)}\right) = \mathbb{E}\left(\exp[-\lambda Z_t(x,0)]s^{P_t(x,0)}\right)$$
$$\times \left[\mathbb{E}\left(\exp[-\lambda Z_t(0,1)]s^{P_t(0,1)}\right)\right]^n. \tag{4.9}$$

Combining the four above identities, we obtain

$$\sum_{n=0}^{\infty}\frac{(x\gamma')^n}{n!}\left[\mathbb{E}\left(\exp[-\lambda Z_t(0,1)]s^{P_t(0,1)}\right)\right]^n$$
$$= \exp\left\{x\left[u(t,\lambda + \gamma') - u(t,\lambda + \gamma'(1-s))\right]\right\}$$
$$= \sum_{n=0}^{\infty}\frac{\{x[u(t,\lambda + \gamma') - u(t,\lambda + \gamma'(1-s))]\}^n}{n!}.$$

Identifying the coefficients of x in the two series yields

$$\mathbb{E}\left(\exp[-\lambda Z_t(0,1)]s^{P_t(0,1)}\right) = \frac{u(t,\lambda + \gamma') - u(t,\lambda + \gamma'(1-s))}{\gamma'}.$$

The result follows from this, (4.9) and (4.8).   □

4.6 A More General Dawson–Li SDE

Because we shall need it below, we study here a generalization of the SDE (4.1), where the drift is nonlinear. Let $f \in C(\mathbb{R}_+)$ satisfy

Assumption (H1) $f(0) = 0$[1] and there exists a constant $\beta \geq 0$ such that for all $x, y \geq 0$, $f(x+y) - f(x) \leq \beta y$.

Remark 7. Here and further in the next chapters, Assumption (H1) could be slightly weakened by assuming

(i) There exists a constant $\beta > 0$ such that $f(x) \leq \beta x$ for all $x > 0$.
(ii) There exists a sequence $(\beta_n, n \geq 1)$ such that $f(x+y) - f(x) \leq \beta_n y$, whenever $x, y \geq 0, x+y \leq n$.

We consider the SDE

$$Z_t^x = x + \int_0^t f(Z_s^x)ds + \sigma \int_0^t \int_0^{Z_s^x} W(ds, du). \tag{4.10}$$

Theorems 1 and 2 are easy to generalize.

Theorem 5. *If f satisfies Assumption (H1), then equation (4.10) has a unique solution. Moreover if Z^x (resp. Z^y) denotes the solution starting from $Z_0^x = x$ (resp. from $Z_0^y = y$) and $x < y$, then $\mathbb{P}(Z_t^x \leq Z_t^y \text{ for all } t > 0) = 1$.*

PROOF: The proof of pathwise uniqueness is an easy extension of that of Theorem 1, thanks to Assumption (H1) (again weak existence follows from Theorem 13 below, and strong existence follows from that and pathwise uniqueness, see section 3.9 in [35]). Indeed, If Z^1 and Z^2 are two solutions, and ϕ_k is as in the proof of Theorem 1, it is easy to check that

$$\phi_k'(Z_s^1 - Z_s^2)[f(Z_s^1) - f(Z_s^2)] \leq \beta|Z_s^1 - Z_s^2|,$$

hence the same estimate as in the above proof holds, with $|\gamma|$ replaced by β.

Concerning the comparison result, we note that with ϕ_k as in the proof of Theorem 2,

$$\phi_k'(Z_s^x - Z_s^y)[f(Z_s^x) - f(Z_s^y)] \leq \beta(Z_s^x - Z_s^y)_+,$$

hence again the same estimate as in the above proof, with $|\gamma|$ replaced by β. $\square$

Remark 8. The comparison result of Theorem 5 can be improved as follows. Suppose $g \in C(\mathbb{R}_+)$ is such that $f(z) \leq g(z)$ for all $z \geq 0$, and again $x \leq y$. If Z_t^x solves the SDE 4.10 and Z_t^y solves

$$Z_t^y = y + \int_0^t g(Z_s^y)ds + \sigma \int_0^t \int_0^{Z_s^y} W(ds, du),$$

then $\mathbb{P}(Z_t^x \leq Z_t^y \text{ for all } t > 0) = 1$.

---

[1] This first assumption is not necessary in Theorem 5 and Lemma 7 below, where it could be replaced by $f(0) \geq 0$, but we shall need it later on.

The proof is very similar to the proof of the comparison result in Theorem 5. Indeed

$$\phi_k'(Z_s^x - Z_s^y)[f(Z_s^x) - g(Z_s^y)] \leq \phi_k'(Z_s^x - Z_s^y)[f(Z_s^x) - f(Z_s^y)]$$
$$\leq \beta(Z_s^x - Z_s^y)_+.$$

$\square$

Clearly the solution of (4.10) is not a branching process, the nonlinear drift introducing interactions between the individuals. In particular, the increments in x are not independent. It follows from Theorem 5 that the mapping $x \to Z^x$ is a.s. increasing from the set of positive rational numbers into $C([0, +\infty); \mathbb{R}_+)$. Let us now establish

Lemma 7. *There exists a constant $C > 0$ such that for any $T > 0$, any $0 < x < y$,*

$$\mathbb{E}\left(\sup_{0 \leq t \leq T} |Z_t^y - Z_t^x|\right) \leq C\left[(y-x)e^{\beta T} + \sqrt{(y-x)e^{\beta T}}\right].$$

PROOF: It is plain that $\inf_{0 \leq t \leq T}(Z_t^y - Z_t^x) \geq 0$ a.s., and there exists a continuous martingale $M^{x,y}$ such that

$$Z_t^y - Z_t^x \leq y - x + \beta \int_0^t (Z_s^y - Z_s^x)ds + M_t^{x,y},$$

$$\langle M^{x,y}\rangle_t = \sigma^2 \int_0^t (Z_s^y - Z_s^x)ds.$$

Hence

$$\mathbb{E}(Z_t^y - Z_t^x) \leq (y-x)e^{\beta t},$$

and, using Doob's inequality on the second line below

$$\sup_{0 \leq s \leq t}(Z_t^y - Z_t^x) \leq y - x + \beta \int_0^t (Z_s^y - Z_s^x)ds + \sup_{0 \leq s \leq t}|M_s^{x,y}|,$$

$$\mathbb{E}\sup_{0 \leq t \leq T}|M_t^{x,y}|^2 \leq c(y-x)e^{\beta T},$$

$$\mathbb{E}\sup_{0 \leq t \leq T}(Z_t^y - Z_t^x) \leq (y-x)e^{\beta T} + \sqrt{c(y-x)e^{\beta T}}.$$

The result follows. $\square$

It follows from Theorem 5 that there exists $N \in \mathscr{F}$ such that $\mathbb{P}(N) = 0$ and for all $\omega \notin N$, $x \mapsto Z^x(\omega)$ is increasing from $\mathbb{Q}_+$ into $C([0, +\infty))$. Whenever $x_n \uparrow x$ or $x_n \downarrow x$ with $x_n, x \in \mathbb{Q}_+$, $Z^{x_n} \to Z^x$ in $C([0, T])$ in probability from Lemma 7, then also a.s. by monotonicity. Hence we can assume that the $\mathbb{P}$-null set N has been chosen in such a way that for $\omega \notin N$, $x \mapsto Z^x(\omega)$ is continuous from $\mathbb{Q}_+$ into $C([0, +\infty))$.

Now choose any $x > 0$, and define

$$\tilde{Z}^x = \lim_{x_n \downarrow x, x_n \in \mathbb{Q}_+, x_n > x} Z^{x_n}.$$

It follows from Lemma 7 that $\tilde{Z}^x = Z^x$ a.s. Hence $\tilde{Z}^x$ solves the SDE (4.10). $\{\tilde{Z}^x_t, t \geq 0, x > 0\}$ is a.s. continuous in t, right-continuous and increasing in x, and it solves the SDE (4.10) for each $x > 0$. From now on we shall consider only that modification of Z^x_t, but drop the $\tilde{}$.

We now want to compare the solution Z^x_t of (4.10) with the CSBP Y^x_t solution of the SDE

$$Y^x_t = x + \beta \int_0^t Y^x_s ds + \sigma \int_0^t \int_0^{Y^x_s} W(ds, du). \qquad (4.11)$$

We know that for any $t > 0$, $x \to Y^x_t$ is right-continuous, has finitely many jumps on any compact interval, and is constant between its jumps. We are going to show that the same is true for Z^x_t. A variant of Theorem 5 would show that, since $f(x) \leq \beta x$, $\mathbb{P}(Z^x_t \leq Y^x_t \text{ for all } t \geq 0) = 1$. We want to establish a stronger comparison result, which holds with a very specific coupling of Z^x_t and Y^x_t.

For each $t > 0$, $x > 0$, let

$$A^x_t(Z) = \cup_{y \leq x, \, y \in D_t} (Y^{y-}_t, Y^{y-}_t + Z^y_t - Z^{y-}_t]$$

$$\text{where } D_t = \{y > 0; \ Y^y_t > Y^{y-}_t\}.$$

Note that the random set A^x_t depends upon the copy of Z, in particular upon the chosen coupling of Y and Z. Note also that the Lebesgue measure of the set $A^x_t(Z)$ equals Z^x_t. We have the

Proposition 13. *There exists a random field $\{\tilde{Z}^x_t, \ x > 0, t \geq 0\}$ such that $t \mapsto \tilde{Z}^x_t$ is continuous, $x \mapsto \tilde{Z}^x_t$ is right-continuous, $\{\tilde{Z}^x_t, \ x > 0, t \geq 0\}$ has the same law as $\{Z^x_t, \ x > 0, t \geq 0\}$ solution of (4.10), $\{\tilde{Z}^x_t, \ x > 0, t \geq 0\}$ solves the SDE*

$$\tilde{Z}^x_t = x + \int_0^t f(\tilde{Z}^x_s) ds + \sigma \int_0^t \int_{A^x_s(\tilde{Z})} W(ds, d\xi), \qquad (4.12)$$

and moreover for all $x, y > 0$,

$$\mathbb{P}(\tilde{Z}^{x+y}_t - \tilde{Z}^x_t \leq Y^{x+y}_t - Y^x_t, \forall t \geq 0) = 1. \qquad (4.13)$$

PROOF: For a solution of (4.12), the equality in law between $\{\tilde{Z}^x_t, \ x > 0, t \geq 0\}$ and $\{Z^x_t, \ x > 0, t \geq 0\}$ follows from the fact that the Lebesgue measure of $A^x_t(\tilde{Z})$ equals $\tilde{Z}^x_t$. We now construct a solution of (4.12).

For each $k, n \geq 1$, let $x^k_n := 2^{-n}k$. For each $n \geq 1$, we now define $\{Z^{n,x}_t, t \geq 0\}$. For $0 < x \leq x^1_n$, we require that $\{Z^{n,x}_t, \ t \geq 0\}$ solves

$$Z^{n,x}_t = x + \int_0^t f(Z^{n,x}_s) ds + \sigma \int_0^t \int_0^{Z^{n,x}_s} W(ds, d\xi).$$

And for $k \geq 2$, we define recursively $\{Z_t^{n,x}, t \geq 0\}$ for $x_n^{k-1} < x \leq x_n^k$ as the solution of

$$Z_t^{n,x} - Z_t^{n,x_n^{k-1}} = x - x_n^{k-1} + \int_0^t \left[f(Z_s^{n,x}) - f(Z_s^{n,x_n^{k-1}}) \right] ds$$
$$+ \sigma \int_0^t \int_{Y_s^{x_n^{k-1}}}^{Y_s^{x_n^{k-1}} + Z_s^{n,x} - Z_s^{n,x_n^{k-1}}} W(ds, d\xi).$$

It is plain that for all $k \geq 1$, $x_n^{k-1} < x \leq x_n^k$,

$$Z_t^{n,x} - Z_t^{n,x_n^{k-1}} \leq Y_t^x - Y_t^{x_n^{k-1}} \quad \text{a.s. for all } t \geq 0, \tag{4.14}$$

and that the law of $\{Z_t^{n,x}, x > 0, t \geq 0\}$ is the same as the law of $\{Z_t^x, x > 0, t \geq 0\}$ solution of (4.10).

Recall that for each $t > 0$, $x \mapsto Y_t^x$ has finitely many jumps on any compact interval, and is constant between its jumps. Moreover for $0 < s < t$,

$$\{x, Y_t^x \neq Y_t^{x-}\} \subset \{x, Y_s^x \neq Y_s^{x-}\}. \tag{4.15}$$

Let us now fix $\delta, M > 0$. For almost any realization of Y, the mapping $x \mapsto Y_\delta^x$ has only finitely many jumps on $(0, M]$. Let n be so large that there is at most one of those jumps in each interval $(k2^{-n}, (k+1)2^{-n}]$, for $k \leq M2^n - 1$. Then for each x that belongs to an interval $(k2^{-n}, (k+1)2^{-n}]$ which contains no jump of $x \mapsto Y_\delta^x$, and for any $n' > n$, we have $Z_t^{n',x} = Z_t^{n,x}$ for any $t \geq \delta$.

Since δ and M are arbitrary positive reals, we have shown that

$$\tilde{Z}_t^x := \text{a.s.} \lim_{n \to \infty} Z_t^{n,x} \tag{4.16}$$

exists for all $t \geq 0$, $x > 0$, the thus constructed random field $\{\tilde{Z}_t^x, t \geq 0, x > 0\}$ has the same law as the solution of the SDE (4.10), and satisfies (4.13) and hence also

$$\{x, \tilde{Z}_t^x \neq \tilde{Z}_t^{x-}\} \subset \{x, Y_t^x \neq Y_t^{x-}\} \tag{4.17}$$

for all $t > 0$. We still have to show that $\tilde{Z}$ satisfies (4.12). It is plain that for any $\delta > 0$,

$$\tilde{Z}_t^x = \tilde{Z}_\delta^x + \int_\delta^t f(\tilde{Z}_s^x) ds + \sigma \int_\delta^t \int_{A_s^x(\tilde{Z})} W(ds, d\xi).$$

In order to deduce that $\tilde{Z}$ satisfies (4.12), it remains to show that $\tilde{Z}_\delta^x \to x$ a.s., as $\delta \to 0$, which follows readily from the equality of the laws of $\tilde{Z}$ and Z. □

We can now deduce

Corollary 2. *For any $t > 0$, $x \mapsto Z_t^x$ has finitely many jumps on any compact interval, and is constant between those jumps.*

PROOF: The assertion follows from the fact that $\tilde{Z}$ possesses that property, as a consequence of (4.17) and the properties of Y. □

The increments of $x \to Z^x$ are not independent, as soon as f is nonlinear. Consider for $0 < x < y$ the increment $V_t^{x,y} := Z_t^y - Z_t^x$. We have

$$V_t^{x,y} = y - x + \int_0^t [f(Z_s^x + V_s^{x,y}) - f(Z_s^x)]ds + \sigma \int_0^t \int_{Z_s^x}^{Z_s^x + V_s^{x,y}} W(ds,du). \quad (4.18)$$

It is plain that $\{Z^{x'}, 0 < x' \leq x\}$ is measurable w.r.t. the sigma field generated by the trace of the white noise $W(dt,du)$ on the random set $A^x := \{(t,u), t \geq 0, 0 \leq u \leq Z_t^x\}$. On the other hand, Z^y is measurable w.r.t. the sigma field generated by $\{Z_t^x, t \geq 0\}$ and the trace of the white noise $W(dt,du)$ on the random set $B^x := \{(t,u), t \geq 0, u > Z_t^x\}$. It follows from the fact that the sets A^x and B^x are disjoint that $\{Z^{x'}, 0 < x' \leq x\}$ and Z^y are conditionally independent, given Z^x. In other words, we have

Proposition 14. $\{Z_t^x, t \geq 0\}_{x \geq 0}$ *is a path-valued Markov process indexed by* x, *which starts from* $Z_t^0 \equiv 0$.

For a description of the generator of this Markov process, we refer the reader to [38].
    We have the

Proposition 15. *The law of the solution* $\{Z_t^x, x > 0, t \geq 0\}$ *of* (4.10) *is uniquely characterized by the following requirements.*
    $Z_t^0 \equiv 0$ *and for all* $0 \leq x < y$, $Z_t^y - Z_t^x \geq 0$ *for all* $t \geq 0$ *a.s.,*

$$M_t^x := Z_t^x - x - \int_0^t f(Z_s^x)ds \quad and$$

$$M_t^{x,y} := Z_t^y - Z_t^x - (y - x) - \int_0^t [f(Z_s^y) - f(Z_s^x)]ds$$

are continuous martingales which satisfy

$$\langle M^x \rangle_t = \sigma^2 \int_0^t Z_s^x ds, \quad \langle M^{xy} \rangle_t = \sigma^2 \int_0^t [Z_s^y - Z_s^x]ds, \quad \langle M^x, M^{x,y} \rangle_t = 0.$$

We leave the proof of this Proposition as an exercise for the reader.
    We now discuss extinction of the process Z_t^x. For $x \geq 0$, define

$$T_0^x = \inf\{t > 0; Z_t^x = 0\}.$$

For any $x \geq 0$, we call the process Z^x (sub)critical (this means either critical or subcritical) if it goes extinct almost surely in finite time, i.e, if T_0^x is finite almost surely. Assumption (H1) implies that $\frac{f(x)}{x}$ is bounded from above. Let us introduce the quantity

$$\Lambda(f) := \int_1^\infty \exp\left(-\frac{2}{\sigma^2} \int_1^u \frac{f(r)}{r}dr\right) du. \quad (4.19)$$

Proposition 16. *Suppose that f satisfies Assumption (H1). For any $x \geq 0$, Z^x is (sub)critical if and only if $\Lambda(f) = \infty$. In particular we have*

i) *A sufficient condition for $\mathbb{P}\left(T_0^x < \infty\right) = 1$ is: there exists $z_0 \geq 1$ such that $f(z) \leq \sigma^2/2, \forall z \geq z_0$,*

ii) *A sufficient condition for $\mathbb{P}\left(T_0^x = \infty\right) > 0$ is: there exists $z_0 > 1$ and $\delta > 0$ such that $f(z) \geq \sigma^2/2 + \delta, \forall z \geq z_0$.*

PROOF: The function

$$S(z) = \int_1^z \exp\left(-\frac{2}{\sigma^2} \int_1^u \frac{f(r)}{r} dr\right) du,$$

which is well defined for any $z > 0$, is a scale function of the diffusion Z^x on $(0, +\infty)$. Let us denote by T_y^x the random time at which Z^x hits y for the first time. We have for any $0 < a < x < b$

$$\mathbb{P}(T_a^x < T_b^x) = \frac{S(b) - S(x)}{S(b) - S(a)}, \text{ and } \mathbb{P}(T_a^x < \infty) = \lim_{b \to \infty} \mathbb{P}(T_a^x < T_b^x).$$

If $\Lambda(f) < \infty$, then $\mathbb{P}(Z_t^x \to \infty$ as $t \to \infty) > 0$, hence $\mathbb{P}(T_0^x < \infty) < 1$, the process is not (sub)critical.

Suppose now that $\Lambda(f) = +\infty$. Since $f(0) = 0$ and f is continuous, we can choose $a > 0$ such that $\delta_a = \sup_{0 \leq x \leq 2a} f(x) < \sigma^2/2$. Our argument shows that Z_t^x gets below a a.s. after any time. But we know that the solution of

$$Y_t = a + \delta_a t + \sigma \int_0^t \sqrt{Y_s} dB_s$$

hits zero before $2a$ with positive probability (see, e.g., Proposition XI.1.5 in [41], where the case $\sigma = 2$ is treated, from which the general case follows easily via a time change). Consequently, thanks to Remark 8, the same is true with Z_t^x after it hits a. The result follows.   □

Remark 9. If we drop the assumption that $f(0) = 0$, then a necessary and sufficient condition for our process Z^x to be (sub)critical is that both $\Lambda(f) = \infty$ and $f(0) < \sigma^2/2$.

Chapter 5
Genealogies

It is not obvious how to trace genealogies for the individuals whose progeny survives
for a given duration of time, also we have seen there is only a finite number of
those, see section 4.4. Another point of view, which we now develop, is to use the
so-called contour or height process. In the case of Feller's diffusion, the bijection
between the contour process and the branching process is given by a Ray–Knight
theorem, see section 5.4 below. Here we shall give an independent derivation of
this theorem, starting with the contour process of a binary continuous time Galton–
Watson process.

There are various forms of bijection between a contour (or height) process and
a random binary tree. This section starts with a description of such a bijection, and
a rather simple proof that a certain law on the contour paths is in bijection with
the law of a continuous time binary Galton–Watson random tree, see Ba, Pardoux,
Sow [6]. The result in the critical case was first established by Le Gall [26], and
in the subcritical case by Pitman and Winkel [39], see also Geiger and Kersting
[19], Lambert [24], where the contour processes are jump processes, while ours
are continuous. For similar results in the case where the approximating process is
in discrete time and the tree is not necessarily binary, see Duquesne and Le Gall
[17]. We consider also the supercritical case. Inspired by the work of Delmas [16],
we note that in the supercritical case, the random tree killed at time $a > 0$ is in
bijection with the contour process reflected below a. Moreover, one can define a
unique local time process, which describes the local times of all the reflected contour
processes, and has the same law as the supercritical Galton–Watson process. We
next renormalize our Galton–Watson tree and height process, and take the weak
limit, thus providing a new proof of Delmas' extension [16] of the second Ray–
Knight theorem. The classical version of this theorem establishes the identity in law
between the local time of reflected Brownian motion considered at the time when the
local time at 0 reaches x, and at all levels, and a Feller critical branching diffusion.
The same result holds in the subcritical (resp. supercritical) case, Brownian motion
being replaced by Brownian motion with drift (in the supercritical case, reflection

© Springer International Publishing Switzerland 2016
É. Pardoux, *Probabilistic Models of Population Evolution*, Mathematical
Biosciences Institute Lecture Series 1.6, DOI 10.1007/978-3-319-30328-4_5

below an arbitrary level, as above, is needed). The contour process in fact describes the genealogical tree (in the sense of Aldous [2]) of the population, whose total mass follows a Feller SDE. Our proof by approximation makes this interpretation completely transparent.

5.1 Preliminaries

The *artificial* time for the evolution of the contour process of our trees will be labelled s, while the *real* time of the evolution of the population is t. t will also label the values taken by the contour process. See the various figures below, where s is always the coordinate of the horizontal axis, and t the coordinate of the vertical axis.

We fix an arbitrary $p > 0$. Consider a continuous piecewise linear function H from a subinterval of $\mathbb{R}_+$ into $\mathbb{R}_+$, which possesses the following properties : its slope is either p or $-p$; it starts at time $s = 0$ from 0 with the slope p; whenever $H(s) = 0, H'_-(s) = -p$, and $H'_+(s) = p$; H is stopped at the time T_m of its m-th return to 0, which is supposed to be finite. We shall denote by $\mathscr{H}_{p,m}$ the collection of all such functions. We shall write $\mathscr{H}_p$ for $\mathscr{H}_{p,1}$. We add the restriction that between two consecutive visits to zero, any function from $\mathscr{H}_{p,m}$ has all its local minima at distinct heights.

We denote by $\mathscr{T}$ the set of continuous time finite rooted binary trees which are defined as follows. An ancestor is born at time 0. This is the root of the tree. Until she eventually dies, she gives birth to an arbitrary number of offsprings, but only one at a time. The same happens to each of her offsprings, and to the offsprings of her offsprings, etc... until eventually the population dies out (assuming for simplicity that we are in the (sub)critical case). For instance, the picture on the right of Figure 5.1 shows a binary tree where the ancestor gives birth to two children before dying. The first child dies childless, while the second one has one child, who dies at the same time as herself. Note that we do not distinguish between the two trees shown in Figure 5.1. We denote by $\mathscr{T}_m$ the set of forests which are the union of m elements of $\mathscr{T}$.

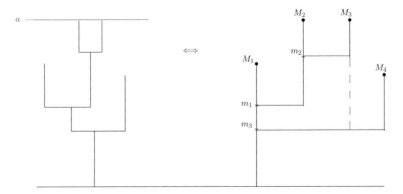

Fig. 5.1 Two equivalent ways of drawing a binary tree

There is a well-known bijection between binary trees and contour processes. Under the curve representing an element $H \in \mathcal{H}_p$, we can draw a tree as follows. The height h_{lfmax} of the leftmost local maximum of H is the lifetime of the ancestor and the height h_{lowmin} of the lowest non zero local minimum is the time of the birth of the first offspring of the ancestor. If there is no such local minimum, the ancestor dies before giving birth to any offspring. We draw a horizontal line at level h_{lowmin}. H has two excursions above h_{lowmin}. The right one is used to represent the fate of the first offspring and of her progeny. The left one is used to represent the fate of the ancestor and of the rest of her progeny, excluding the first offspring and her progeny. If there is no other local minimum of H neither at the left nor at the right of the first explored one, it means that there is no further birth: we draw a vertical line up to the unique local maximum, whose height is a death time. Continuing until there is no further local minimum-maximum to explore, we define by this procedure a bijection Φ_p from $\mathcal{H}_p$ into $\mathcal{T}$ (see Figure 5.2). Repeating the same construction m times, we extend Φ_p as a bijection between $\mathcal{H}_{p,m}$ and $\mathcal{T}_m$. Note that drawing the contour process of a tree is obvious (since the horizontal distances between the vertical branches are arbitrary, the choice of p is arbitrary). See the top of Figure 5.2.

We now define probability measures on $\mathcal{H}_p$ (resp. $\mathcal{H}_{p,m}$) and $\mathcal{T}$ (resp. $\mathcal{T}_m$). We describe first the (sub)critical case. Let $0 < b \leq d$ be two parameters. We define a stochastic process whose trajectories belong to $\mathcal{H}_p$ as follows. Let $\{U_k, k \geq 1\}$ and $\{V_k, k \geq 1\}$ be two mutually independent sequences of i.i.d exponential random variables with means $1/d$ and $1/b$, respectively. Let $Z_k = U_k - V_k, k \geq 1$. $\mathbb{P}_{b,d}$ is the law of the random element of $\mathcal{H}_p$, which is such that the height of the first local maximum is U_1, that of the first local minimum is $(Z_1)^+$. If $(Z_1)^+ = 0$, the process is stopped. Otherwise, the height of the second local maximum is $Z_1 + U_2$, the height of the second local minimum is $(Z_1 + Z_2)^+$, etc. Because $b \leq d$, $\mathbb{E}Z_1 \leq 0$, hence the process returns to zero a.s. in finite time. The random trajectory which we have constructed is an excursion above zero (see the bottom of Figure 5.2). We define similarly a law on $\mathcal{H}_{p,m}$ as the concatenation of m i.i.d. such excursions, and denote it by $\mathbb{P}_{b,d}$. This thus defined random element of $\mathcal{H}_{p,m}$ is called a contour or height process. We associate the continuous time Galton–Watson tree (which is a random element of $\mathcal{T}$) with the same pair of parameters (b,d) as follows. The lifetime of each individual is exponential with expectation $1/d$, and during her lifetime, independently of it, each individual gives birth to offsprings according to a Poisson process with rate b. The behaviors of the various individuals are i.i.d. We denote by $\mathbb{Q}_{b,d}$ the law on $\mathcal{T}_m$ of a forest of m i.i.d. random trees whose law is as just described.

In the supercritical case, the case where $b > d$, the contour process defined above does not come back to zero a.s.. To overcome this difficulty, we use a trick which is due to Delmas [16], and reflect the process H below an arbitrary level $a > 0$ (which amounts to kill the whole population at time a). The height process $H^a = \{H_s^a, s \geq 0\}$ reflected below a is defined as above, with the addition of the rule that whenever the process reaches the level a, it stops and starts immediately going down with slope $-p$ for a duration of time exponential with expectation $1/b$. Again the

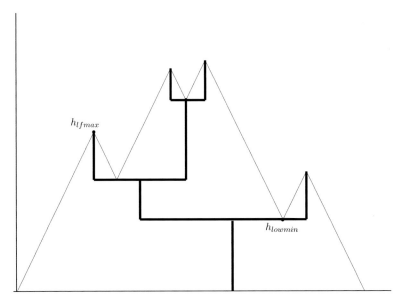

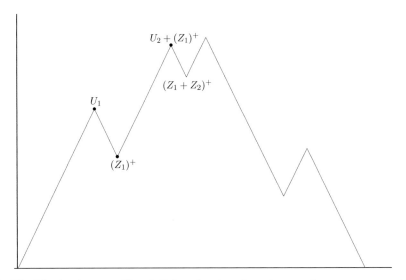

Fig. 5.2 Bijection between $\mathscr{H}_2$ and $\mathscr{T}$ (see above), and a trajectory of a contour process (see below)

process stops when first going back to zero. The reflected process H^a comes back to zero almost surely. Indeed, let A_n^a denote the event "H^a does not reach zero during its n first descents." Since the levels of the local maxima are bounded by a, clearly $\mathbb{P}(A_n) \leq (1 - \exp(-ba))^n$, which goes to zero as $n \longrightarrow \infty$. Hence the result. For each $a \in (0, +\infty)$, and any pair (b,d) of positive numbers, denote by $\mathbb{P}_{b,d,a}$ the law of the process H^a. Define $\mathbb{Q}_{b,d,a}$ to be the law of a binary Galton–Watson tree with birth rate b and death rate d, killed at time $t = a$], (i. e. all individuals alive at time a^- are killed at time a). $\mathbb{P}_{b,d,+\infty}$ makes perfect sense in case $b \leq d$, $\mathbb{Q}_{b,d,+\infty}$ is always well defined.

5.2 Correspondence of Laws

The aim of this section is to prove that, for any $b,d > 0$ and $a \in (0, +\infty)$ [including possibly $a = +\infty$ in the case $b \leq d$], $\mathbb{P}_{b,d,a}\Phi_p^{-1} = \mathbb{Q}_{b,d,a}$. Let us state some basic results on homogeneous Poisson process, which will be useful in the sequel.

5.2.1 Preliminary Results

Let $(T_k)_{k \geq 0}$ be a Poisson point process on $\mathbb{R}_+$ with intensity (or rate) b. This means that $T_0 = 0$, and $(T_{k+1} - T_k, \, k \geq 0)$ are i.i.d exponential r.v.'s with mean $1/b$. Let $(N_t, \, t \geq 0)$ be the counting process associated with $(T_k)_{k \geq 0}$, that is,

$$\forall t \geq 0, \, N_t = \sup\{k \geq 0, \, T_k \leq t\}.$$

We shall also call $(N_t, \, t \geq 0)$ a Poisson process. This process has independent increments, and for any $0 \leq s < t$, $N_t - N_s$, which is the number of points of the above PPP in the interval $(s,t]$, follows the Poisson law with parameter $b(t - s)$. In particular, the intensity b is the mean number of points of the PPP in an interval of length 1.

The first result is well-known and elementary.

Lemma 8. *Let M be a nonnegative random variable independent of $(T_k)_{k \geq 0}$, and define*

$$R_M = \sup_{k \geq 0}\{T_k; T_k \leq M\}. \tag{5.1}$$

Then $M - R_M \overset{(d)}{=} V \wedge M$ where V and M are independent, V has an exponential distribution with mean $1/b$.
Moreover, on the event $\{R_M > s\}$, the conditional law of $N_{R_M^-} - N_s$ given R_M is Poisson with parameter $b(R_M - s)$.

The second one is (in the next statement, we use again the definition (5.1)) :

Lemma 9. *Let $(T_k)_{k \geq 0}$ be a Poisson point process on $\mathbb{R}_+$ with intensity b. M a positive random variable which is independent of $(T_k)_{k \geq 0}$. Consider the integer-valued*

random variable K such that $T_K = R_M$ and a second Poisson point process $(T'_k)_{k \geq 0}$
with intensity b, which is jointly independent of the first and of M. Then $(\overline{T}_k)_{k \geq 0}$
defined by:

$$\overline{T}_k = \begin{cases} T_k & \text{if } k < K \\ T_K + T'_{k-K+1} & \text{if } k \geq K \end{cases}$$

is a Poisson point process on $\mathbb{R}_+$ with intensity b, which is independent of R_M.

PROOF: Let $(N_t, t \geq 0)$, $(\overline{N}_t, t \geq 0)$ and $(N'_t, t \geq 0)$) be the counting processes associated to $(T_k)_{k \geq 0}$, $(\overline{T}_k)_{k \geq 0}$ and $(T'_k)_{k \geq 0}$, respectively. It suffices to prove that for any $n \geq 1$, $0 < t_1 < \cdots < t_n$ and $k_1 < k_2 < \cdots < k_n \in \mathbb{N}^*$,

$$\xi_M = \mathbb{P}\left(\overline{N}_{t_1} = k_1, \ldots, \overline{N}_{t_n} = k_n | R_M \right) = e^{-bt_n} \prod_{i=1}^{n} \frac{(b(t_i - t_{i-1}))^{k_i - k_{i-1}}}{(k_i - k_{i-1})!}.$$

Since there is no harm in adding $t'_i s$, we only need to do that computation on the event that there exists $2 \leq i \leq n$ such that $t_{i-1} < R_M < t_i$, in which case a standard argument yields easily the claimed result, thanks to Lemma 8. Indeed we have that

$$\xi_M = \mathbb{P}\left(N_{t_1} = k_1, \cdots, N_{t_{i-1}} = k_{i-1}, N_{R_M^-} + N'_{t_i - R_M} = k_i, \cdots, N_{R_M^-} + N'_{t_n - R_M} = k_n \right)$$

$$= \mathbb{P}\left(N_{t_1} = k_1, \cdots, N_{t_{i-1}} - N_{t_{i-2}} = k_{i-1} - k_{i-2}, N_{R_M^-} - N_{t_{i-1}} + N'_{t_i - R_M} = k_i - k_{i-1}, \right.$$

$$\left. N'_{t_{i+1} - R_M} - N'_{t_i - R_M} = k_{i+1} - k_i, \cdots, N'_{t_n - R_M} - N'_{t_{n-1} - R_M} = k_n - k_{n-1} \right)$$

$$= e^{-bt_n} \prod_{i=1}^{n} \frac{(b(t_i - t_{i-1}))^{k_i - k_{i-1}}}{(k_i - k_{i-1})!}.$$

$\square$

5.2.2 Basic Theorem

We are now in a position to prove the next theorem, which says that the tree associated to the contour process H^a defined in section 5.1 is a continuous time binary Galton–Watson tree with death rate d and birth rate b, killed at time a, and vice versa.

Theorem 6. *For any $b,d > 0$ and $a \in (0,+\infty)$ [including possibly $a = +\infty$ in the case $b \le d$],*

$$\mathbb{Q}_{b,d,a} = \mathbb{P}_{b,d,a}\Phi_p^{-1}.$$

PROOF: The individuals making up the population represented by the tree whose law is $\mathbb{Q}_{b,d,a}$, will be numbered: $\ell = 1,2,\ldots$. 1 is the ancestor of the whole family. The subsequent individuals will be identified below. We will show that this tree is explored by a process whose law is precisely $\mathbb{P}_{b,d,a}$. We introduce a family $(T_k^\ell, k \ge 0, \ell \ge 1)$ of mutually independent Poisson point processes with intensity b. For any $\ell \ge 1$, the process T_k^ℓ describes the times of birth of the offsprings of the individual ℓ. We define U_ℓ to be the lifetime of individual ℓ.

- **Step 1**: We start from $H_0^a = 0$ at the initial time $s = 0$ and climb up with slope p to the level $M_1 = U_1 \wedge a$, where U_1 follows an exponential law with mean $1/d$. H_s^a goes down from M_1 with slope $-p$ until we find the most recent point of the Poisson process (T_k^1) which gives the times of birth of the offsprings of individual 1. So from Lemma 8, H_s^a has descended by $V_1 \wedge M_1$, where V_1 follows an exponential law with mean $1/b$, and is independent of M_1. We hence reach the level $m_1 = M_1 - V_1 \wedge M_1$.
  If $m_1 = 0$, we stop, else we turn to
- **Step 2**: We give the label 2 to this last offspring of the individual 1, born at the time m_1. Let us define $(\bar{T}_k^2)$ by:

$$\bar{T}_k^2 = \begin{cases} T_k^1 & \text{if } k < K_1 \\ T_{K_1}^1 + T_{k-K_1+1}^2 & \text{otherwise} \end{cases}$$

where K_1 is such that $T_{K_1}^1 = m_1$.
Thanks to Lemma 9, $(\bar{T}_k^2)$ is a Poisson process with intensity b on $\mathbb{R}_+$, which is independent of m_1 and in fact also of (U_1,V_1).
Starting from m_1, the contour process climbs up to level $M_2 = (m_1 + U_2) \wedge a$, where U_2 is an exponential r.v. with mean $1/d$, independent of (U_1,V_1). Starting from level M_2, we go down a height $M_2 \wedge V_2$ where V_2 follows an exponential law with mean $1/b$ and is independent of (U_2,U_1,V_1), to find the most recent point of the Poisson process $(\bar{T}_k^2)$. At this moment we are at the level $m_2 = M_2 - V_2 \wedge M_2$. If $m_2 = 0$ we stop. Otherwise we give the label 3 to the individual born at time m_2, and repeat step 2 until we reach 0. See Figure 5.1.

Since either we have a reflection at level a or $b \le d$, zero is reached a.s. after a finite number of iterations. It is clear that the random variables M_i and m_i determine fully the law $\mathbb{Q}_{b,d,a}$ of the binary tree killed at time $t = a$ and they both have the same joint distribution as the levels of the successive local maxima and minima of the process H^a under $\mathbb{P}_{b,d,a}$. $\qquad\square$

5.2.3 A Discrete Ray–Knight Theorem

For any $a,b,d > 0$, we consider the contour process $\{H_s^a,\ s \geq 0\}$ defined in section 5.1 which is reflected in the interval $[0,a]$ and stopped at the first moment it reaches zero for the m-th time. To this process we can associate a forest of m binary trees with birth rate b and death rate d, killed at time $t = a$, which all start with a single individual at the initial time $t = 0$. Consider the continuous time branching process $\left(X_t^{a,m},\ t \geq 0\right)$ describing the number of offsprings alive at time t of the m ancestors born at time 0, whose progeny is killed at time $t = a$. Every individual in this population, independently of the others, lives for an exponential time with parameter d and gives birth to offsprings according to a Poisson process of intensity b. We now choose the slopes of the piecewise linear process H^a to be ± 2 (i.e., $p = 2$). We define the local time accumulated by H^a at level t up to time s:

$$L_s^a(t) = \lim_{\varepsilon \downarrow 0} \frac{1}{\varepsilon} \int_0^s \mathbf{1}_{\{t \leq H_r^a < t + \varepsilon\}} dr. \qquad (5.2)$$

$L_s^a(t)$ equals the number of pairs of branches of H^a which cross level t between times 0 and s. Note that a local minimum at level t counts for two crossings, while a local maximum at level t counts for none. We have the following "occupation times formula," whose proof is an easy exercise. For any integrable function g,

$$\int_0^s g(H_r^a) dr = \int_0^a g(r) L_s^a(r) dr. \qquad (5.3)$$

Let

$$\tau_m^a = \inf\left\{s > 0 : L_s^a(0) \geq m\right\}. \qquad (5.4)$$

$L_{\tau_m^a}^a(t)$ counts the number of descendants of m ancestors at time 0, which are alive at time t. Then we have

Lemma 10. *For all $b,d > 0$ and $a \in (0,+\infty)$ [including possibly $a = +\infty$ in the case $b \leq d$].*

$$\left\{L_{\tau_m^a}^a(t),\ t \geq 0, m \geq 1\right\} \equiv \left\{X_t^{a,m},\ t \geq 0, m \geq 1\right\} \ a.s.$$

We now want to establish a similar statement without the arbitrary parameter a. There remains a difficulty only in the supercritical case, in which case we cannot choose $a = +\infty$ in the above construction. For any $0 < a < b$, we define the function $\Pi^{a,b}$ which maps continuous trajectories with values in $[0,b]$ into trajectories with values in $[0,a]$ as follows. If $u \in C(\mathbb{R}_+, [0,b])$,

$$\rho_u(s) = \int_0^s \mathbf{1}_{\{u(r) \leq a\}} dr; \quad \Pi^{a,b}(u)(s) = u(\rho_u^{-1}(s)). \qquad (5.5)$$

Lemma 11.

$$\Pi^{a,b}(H^b) \overset{(d)}{=} H^a$$

PROOF: It is in fact sufficient to show that the conditional law of the level of the first local minimum of H^b after crossing the level a downwards, given the past of H^b, is the same as the conditional law of the level of the first local minimum of H^a after a reflection at level a, given the past of H^a. This identity follows readily from the "lack of memory" of the exponential law. $\quad\square$

This last Lemma says that reflecting under a, or chopping out the pieces of trajectory above level a, yields the same result (at least in law).

We now consider the case $p = 2$. To each $b, d > 0$, $m \geq 1$, we associate the process $\{X_t^m, \, t \geq 0\}$ which describes the evolution of the number of descendants of m ancestors, with birth rate b and death rate d. For each $a > 0$ [including possibly $a = +\infty$ in the case $b \leq d$], we let $(H_s^a, \, s \geq 0)$ denote the contour process of the genealogical tree of this population killed at time a, L^a denotes its local time and τ_m^a is defined by (5.4). It follows readily from Lemma 11 that for any $0 < a < b$,

$$\left(L_{\tau_m^b}^b(t), 0 \leq t < a, m \geq 1 \right) \overset{(d)}{=} \left(L_{\tau_m^a}^a(t), 0 \leq t < a, m \geq 1 \right). \tag{5.6}$$

The compatibility relation (5.6) implies the existence of a projective limit $\{\mathscr{L}_m(t), t \geq 0, m \geq 1\}$ with values in $\mathbb{R}_+$, which is such that for each $a > 0$,

$$\{\mathscr{L}_m(t), 0 \leq t < a, m \geq 1\} \overset{(d)}{=} \{L_{\tau_m^a}^a(t), 0 \leq t < a, m \geq 1\}. \tag{5.7}$$

We have the following "discrete Ray–Knight Theorem."

Proposition 17.

$$\{\mathscr{L}_m(t), t \geq 0, m \geq 1\} \overset{(d)}{=} \{X_t^m, \, t \geq 0, m \geq 1\}.$$

PROOF: It suffices to show that for any $a \geq 0$,

$$\{\mathscr{L}_m(t), 0 \leq t < a, m \geq 1\} \overset{(d)}{=} \{X_t^m, 0 \leq t < a, m \geq 1\}.$$

This result follows from (5.7) and Lemma 10. $\quad\square$

5.2.4 Renormalization

Let $x > 0$ be arbitrary, and $N \geq 1$ be an integer which will eventually go to infinity. Let $(X_t^{N,x})_{t \geq 0}$ denote the branching process which describes the number of descendants at time t of $[Nx]$ ancestors, in the population with birth rate $b_N = \sigma^2 N/2 + \alpha$ and death rate $d_N = \sigma^2 N/2 + \beta$, where $\alpha, \beta \geq 0$. We set for $t \geq 0$

$$Z_t^{N,x} = N^{-1} X_t^{N,x}.$$

In particular we have that $Z_0^{N,x} = \frac{[Nx]}{N} \to x$ when $N \to +\infty$. Let $H^{a,N}$ be the contour process associated to $\{X_t^{N,x}, 0 \le t < a\}$ defined in the same way as previously, but with slopes $\pm 2N$, and b and d are replaced by b_N and d_N. We define also $L_s^{a,N}(t)$, the local time accumulated by $H^{a,N}$ at level t up to time s, as

$$L_s^{a,N}(t) = \frac{4}{\sigma^2} \lim_{\varepsilon \downarrow 0} \frac{1}{\varepsilon} \int_0^s \mathbf{1}_{\{t \le H_r^{a,N} < t+\varepsilon\}} dr \qquad (5.8)$$

$\frac{\sigma^2}{4} L_s^{a,N}(t)$ equals $1/N$ times the number of pairs of t-crossings of $H^{a,N}$ between times 0 and s. Let

$$\tau_x^{a,N} = \inf\left\{ s > 0 : L_s^{a,N}(0) \ge \frac{4}{\sigma^2} \frac{[Nx]}{N} \right\}. \qquad (5.9)$$

We define for all $N \ge 1$ the projective limit $\{\mathcal{L}_x^N(t), t \ge 0, x > 0\}$, which is such that for each $a > 0$,

$$\{\mathcal{L}_x^N(t), 0 \le t < a, x > 0\} \overset{(d)}{=} \{L_{\tau_x^{a,N}}^{a,N}(t), 0 \le t < a, x > 0\}.$$

Proposition 17 translates as (note that the factor N^{-1} in the definition of $Z_t^{N,x}$ matches the slopes $\pm 2N$ of $H^{a,N}$, which introduces a factor N^{-1} in the local times defined by (5.8))

Lemma 12. *We have the identity in law*

$$\{\mathcal{L}_x^N(t), t \ge 0, x > 0\} \overset{(d)}{=} \left\{ \frac{4}{\sigma^2} Z_t^{N,x}, t \ge 0, x > 0 \right\}.$$

We will need to write precisely the evolution of $\{H^{a,N}, s \ge 0\}$, the contour process of the forest of trees representing the population $\{X_t^{N,x}, 0 \le t < a\}$. Let $\{V_s^{a,N}, s \ge 0\}$ be the $\{-1,1\}$-valued process which is such that s-a.e. $\frac{dH_s^{a,N}}{ds} = 2NV_s^{a,N}$. The $\mathbb{R}_+ \times \{-1,1\}$-valued process $\{(H_s^{a,N}, V_s^{a,N}), s \ge 0\}$ is a Markov process, which solves the following SDE :

$$\frac{dH_s^{a,N}}{ds} = 2NV_s^{a,N}, \quad H_0^{a,N} = 0, V_0^{a,N} = 1,$$

$$dV_s^{a,N} = 2\mathbf{1}_{\{V_{s-}^{a,N}=-1\}} dP_s^+ - 2\mathbf{1}_{\{V_{s-}^{a,N}=1\}} dP_s^- + \frac{\sigma^2}{2} N dL_s^{a,N}(0) - \frac{\sigma^2}{2} N dL_s^{a,N}(a^-),$$
$$(5.10)$$

where $\{P_s^+, s \ge 0\}$ and $\{P_s^-, s \ge 0\}$ are two mutually independent Poisson processes, with intensities (given by $2N \times$ the rate of birth (resp. death))

$$\sigma^2 N^2 + 2\alpha N \quad \text{and} \quad \sigma^2 N^2 + 2\beta N,$$

$L_s^{a,N}(0)$ and $L_s^{a,N}(a^-)$ denote, respectively, the number of visits to 0 and a by the process $H^{a,N}$ up to time s, multiplied by $4/N\sigma^2$ (see (5.8)). These two terms in the expression of $V^{a,N}$ stand for the reflection of $H^{a,N}$ above 0 and below a. Note that our definition of $L^{a,N}$ makes the mapping $t \longrightarrow L_s^{a,N}(t)$ right-continuous for each $s > 0$. Hence $L_s^{a,N}(t) = 0$ for $t \geq a$, while $L_s^{a,N}(a^-) = \lim_{n\to\infty} L_s^{a,N}(a - \frac{1}{n}) > 0$ if $H^{a,N}$ has reached the level a by time s.

5.3 Weak Convergence

5.3.1 Tightness of $H^{a,N}$

We deduce from (5.10) that

$$H_s^{a,N} = 2N \int_0^s \mathbf{1}_{\{V_r^{a,N}=1\}} dr - 2N \int_0^s \mathbf{1}_{\{V_r^{a,N}=-1\}} dr,$$

$$\frac{V_s^{a,N}}{\sigma^2 N} = \frac{1}{\sigma^2 N} - \left(2N + \frac{4\beta}{\sigma^2}\right) \int_0^s \mathbf{1}_{\{V_r^{a,N}=1\}} dr + \left(2N + \frac{4\alpha}{\sigma^2}\right) \int_0^s \mathbf{1}_{\{V_r^{a,N}=-1\}} dr$$

$$- \frac{2}{\sigma^2 N} \int_0^s \mathbf{1}_{\{V_{r^-}^{a,N}=1\}} dM_r^- + \frac{2}{\sigma^2 N} \int_0^s \mathbf{1}_{\{V_{r^-}^{a,N}=-1\}} dM_r^+$$

$$+ \frac{1}{2}[L_s^{a,N}(0) - L_{0^+}^{a,N}(0)] - \frac{1}{2}L_s^{a,N}(a^-),$$

where

$$M_s^+ = P_s^+ - (\sigma^2 N^2 + 2\alpha N)s, \quad M_s^- = P_s^- - (\sigma^2 N^2 + 2\beta N)s$$

are two martingales. Consequently

$$H_s^{a,N} + \frac{V_s^{a,N}}{\sigma^2 N} = \frac{1}{\sigma^2 N} + \frac{4}{\sigma^2} \int_0^s \left(\alpha \mathbf{1}_{\{V_r^{a,N}=-1\}} - \beta \mathbf{1}_{\{V_r^{a,N}=1\}}\right) dr + M_s^{a,N}$$

$$+ \frac{1}{2}[L_s^{a,N}(0) - L_{0^+}^{a,N}(0)] - \frac{1}{2}L_s^{a,N}(a^-),$$

(5.11)

where $M_s^{a,N}$ is a martingale such that

$$[M^{a,N}]_s = \frac{4}{\sigma^4 N^2} \left(\int_0^s \mathbf{1}_{\{V_{r^-}^{a,N}=1\}} dP_r^- + \int_0^s \mathbf{1}_{\{V_{r^-}^{a,N}=-1\}} dP_r^+\right),$$

$$\langle M^{a,N}\rangle_s = \frac{4}{\sigma^2} s + \frac{8}{\sigma^4 N} \int_0^s \left(\alpha \mathbf{1}_{\{V_r^{a,N}=-1\}} + \beta \mathbf{1}_{\{V_r^{a,N}=1\}}\right) dr.$$

Lemma 13. *For any $a > 0$, the sequence $\{H_s^{a,N}, s \geq 0\}_{N \geq 1}$ is tight in $C([0,\infty))$.*

PROOF: Let us rewrite (5.11) in the form

$$H_s^{a,N} = K_s^{a,N} + \frac{1}{2}[L_s^{a,N}(0) - L_{0+}^{a,N}(0)] - \frac{1}{2}L_s^{a,N}(a^-).$$

It follows readily from Proposition 37 below that $K^{a,N}$ is tight in $D([0,\infty))$, and all its jumps converge to 0 as $N \to \infty$. It then follows from Theorem 13.2 and (12.9) in [10] that for any $T > 0$, all $\varepsilon, \eta > 0$, there exist N_0 and $\delta > 0$ such that for all $N \ge N_0$,

$$\mathbb{P}\left(\sup_{0 \le r, s \le T,\ |s-r| \le \delta} |K_s^{a,N} - K_r^{a,N}| > \varepsilon\right) \le \eta.$$

Since $L^N(0)$ (resp. $L^N(a^-)$) increases only when $H_s^N = 0$ (resp. when $H_s^N = a$), it is not hard to conclude that, provided $\varepsilon < a$, the above implies that for all $N \ge N_0$,

$$\mathbb{P}\left(\sup_{0 \le r, s \le T,\ |s-r| \le \delta} |H_s^{a,N} - H_r^{a,N}| > \varepsilon\right) \le \eta.$$

In view of (12.7) in [10], this implies that $H^{a,N}$ is tight in $D([0,\infty))$. $\square$

5.3.2 Weak Convergence of $H^{a,N}$

Let us state our convergence result.

Theorem 7. *For any $a > 0$ [including possibly $a = +\infty$ in the case $\alpha \le \beta$], $H^{a,N} \Rightarrow H^a$ in $C([0,\infty))$ as $N \to \infty$, where $\{H_s^a,\ s \ge 0\}$ is the process*

$$\frac{2(\alpha - \beta)}{\sigma^2}s + \frac{2}{\sigma}B_s$$

reflected in $[0,a]$. In other words, H^a is the unique weak solution of the reflected SDE[1]

$$H_s^a = \frac{2(\alpha - \beta)}{\sigma^2}s + \frac{2}{\sigma}B_s + \frac{1}{2}L_s^a(0) - \frac{1}{2}L_s^a(a^-), \qquad (5.12)$$

where $L_s^a(t)$ denotes the local time accumulated by $(H_r^a,\ r \ge 0)$ up to time s at level t.

We first prove

Lemma 14. *For any sequence $(U^N, N \ge 1) \subset C([0,+\infty))$ which is such that $U^N \Rightarrow U$ as $N \to \infty$, for all $s > 0$,*

$$\int_0^s \mathbf{1}_{\{V_r^{a,N}=1\}} U_r^N\, dr \Rightarrow \frac{1}{2}\int_0^s U_r\, dr, \qquad \int_0^s \mathbf{1}_{\{V_r^{a,N}=-1\}} U_r^N\, dr \Rightarrow \frac{1}{2}\int_0^s U_r\, dr.$$

---

[1] The fact that Brownian motion with drift reflected in the interval $[0,a]$ takes this form is explained at the end of section A.4 below.

PROOF: It is an easy exercise to check that the mapping

$$\Phi : C([0,+\infty)) \times C_\uparrow([0,+\infty)) \to C([0,+\infty))$$

defined by

$$\Phi(x,y)(t) = \int_0^t x(s)dy(s),$$

where $C_\uparrow([0,+\infty))$ denotes the set of increasing continuous functions from $[0,\infty)$ into $\mathbb{R}$, and the three spaces are equipped with the topology of locally uniform convergence, is continuous. Consequently it suffices to prove that locally uniformly in $s > 0$,

$$\int_0^s \mathbf{1}_{\{V_r^{a,N}=1\}} dr \to \frac{s}{2}$$

in probability, as $N \to \infty$. In fact since both the sequence of processes and the limit are continuous and monotone, it follows from an argument "à la Dini" that it suffices to prove

Lemma 15. *For any $s > 0$,*

$$\int_0^s \mathbf{1}_{\{V_r^{a,N}=1\}} dr \to \frac{s}{2}, \quad \int_0^s \mathbf{1}_{\{V_r^{a,N}=-1\}} dr \to \frac{s}{2}$$

in probability, as $N \to \infty$.

PROOF: We have (the second line follows from (5.10))

$$\int_0^s \mathbf{1}_{\{V_r^{a,N}=1\}} dr + \int_0^s \mathbf{1}_{\{V_r^{a,N}=-1\}} dr = s,$$

$$\int_0^s \mathbf{1}_{\{V_r^{a,N}=1\}} dr - \int_0^s \mathbf{1}_{\{V_r^{a,N}=-1\}} dr = (2N)^{-1} H_s^{a,N}.$$

It follows readily from Lemma 13 that $(2N)^{-1}H_s^{a,N} \to 0$ in probability, as $n \to \infty$. We conclude by adding and subtracting the two above identities. □

PROOF OF THEOREM 7 Let us prove that

$$\left(H^{a,N}, M^{N,a}, L_\cdot^{a,N}(0), L_\cdot^{a,N}(a^-) \right) \Longrightarrow \left(H^a, \frac{2}{\sigma}B, L_\cdot^a(0), L_\cdot^a(a^-) \right).$$

Concerning tightness, we only need to take care of the third and fourth terms in the quadruple. We notice that Lemma 13 implies that $(L^{a,N}(0) - L^N(a^-))_{N\geq1}$ is tight in $D([0,\infty))$. Since $L^N(0)$ (resp. $L^N(a^-)$) increases only when $H_s^N = 0$ (resp. when $H_s^N = a$), it is not hard to deduce that both $(L^{a,N}(0))_{N\geq1}$ and $(L^N(a^-))_{N\geq1}$ are tight in $D([0,\infty))$. Alternatively tightness of $L^N(0)$ can be deduced from the identity (5.13) below, and that of $L^N(a^-)$ from a similar formula.

Then $\left(H^{a,N}, M^{N,a}, L_\cdot^{a,N}(0), L_\cdot^{a,N}(a^-) \right)_{N\geq1}$ is tight in $C([0,\infty)) \times [D([0,\infty))]^3$. Moreover any weak limit of $M^{N,a}$ along a converging subsequence equals $2B/\sigma$, since $< M^{N,a} >_s \to 4s/\sigma^2$ and the jumps of $M^{N,a}$ are equal in amplitude to $\frac{2}{\sigma^2 N}$.

Let $f \in C^2(\mathbb{R})$ such that $f'(0) = 1$ and $f'(a) = 0$, and define $f^N(h, v) = f(h) + \frac{v}{\sigma^2 N} f'(h)$. We deduce from (5.10)

$$L_s^{a,N}(0) = 2f(H_s^{a,N}) + \frac{2V_s^{a,N}}{\sigma^2 N} f'(H_s^{a,N}) - 2f(0) - \frac{2}{\sigma^2 N} f'(0) - \frac{4}{\sigma^2} \int_0^s f''(H_r^{a,N}) dr$$

$$- \frac{8}{\sigma^2} \int_0^s f'(H_r^{a,N})(\alpha \mathbf{1}_{\{V_r^N = -1\}} - \beta \mathbf{1}_{\{V_r^N = 1\}}) dr - 2M_s^{f,N} - 2\tilde{M}_s^{f,N},$$

(5.13)

where $M^{f,N}$ and $\tilde{M}^{f,N}$ are martingales such that

$$\langle M^{f,N} \rangle_s = \frac{4}{\sigma^2} \int_0^s [f'(H_r^{a,N})]^2 dr, \quad \langle \tilde{M}^{f,N} \rangle_s \leq \frac{c(f)}{N} s.$$

It follows by taking the limit in (5.13) (and in a similar formula for $L_s^{a,N}(a^-)$) that we have a limit of the form $(H^a, 2B/\sigma, L^a(0), L^a(a^-))$ along a converging subsequence of the sequence $(H^{a,N}, M^{N,a}, L^{a,N}(0), L^{a,N}(a^-))$.

For any $0 < \varepsilon < a$, let $f_\varepsilon \in C^2(\mathbb{R})$ be such that $f_\varepsilon'(0) = 1$, $f_\varepsilon'(x) = 0$ for all $\varepsilon \leq x \leq a$, $f'(x) \geq 0$ and $f''(x) \leq 0$ for all $x \geq 0$. Taking the limit along the converging subsequence in (5.13) with $f_\varepsilon^N(h, v) = f_\varepsilon(h) + \frac{v}{\sigma^2 N} f_\varepsilon'(h)$, we deduce that

$$L_s^a(0) = 2f_\varepsilon(H_s^a) - 2f_\varepsilon(0) - \frac{4}{\sigma^2} \int_0^s f_\varepsilon''(H_r^a) dr - \frac{4}{\sigma^2}(\alpha - \beta) \int_0^s f_\varepsilon'(H_r^a) dr - 2M_s^{f_\varepsilon},$$

$$\langle M^{f_\varepsilon} \rangle_s = \frac{4}{\sigma^2} \int_0^s [f_\varepsilon'(H_r^a)]^2 dr.$$

It then follows that $\int_0^s \mathbf{1}_{\{H_r^a \geq \varepsilon\}} dL_r^a(0) = 0$. This being true for all $0 < \varepsilon < a$, we have that $L_s^a(0) = \int_0^s \mathbf{1}_{\{H_r^a = 0\}} dL_r^a(0)$. We prove similarly that $L_s^a(a^-) = \int_0^s \mathbf{1}_{\{H_r^a = a\}} dL_r^a(a^-)$. Moreover it is plain that both $L_s^a(0)$ and $L_s^a(a^-)$ are continuous and increasing. Now (5.12) follows by taking the limit in (5.11). It is plain that H^a, being a limit (along a subsequence) of $H^{a,N}$, takes values in $[0, a]$. The fact that $L^a(0)$ (resp. $L^a(a^-)$) is continuous and increasing, and increases only on the set of time when $H_s^a = 0$ (resp. $H_s^a = a$) proves that $\frac{\sigma}{2} H^a$ is a Brownian motion with drift $(\alpha - \beta)s/\sigma$, reflected in $[0, a]$, which characterizes its law. We can refer, e.g., to the formulation of reflected SDEs in [30]. Hence the whole sequence converges, and the Theorem is proved. $\square$

We have proved in particular

Corollary 3. *For each $a > 0$ (including $a = +\infty$ in the case $\alpha \leq \beta$),*

$$\left(H^{a,N}, M^{N,a}, L^{a,N}(0), L^{a,N}(a^-)\right) \Rightarrow \left(H^a, \frac{2}{\sigma} B, L^a(0), L^a(a^-)\right)$$

as $N \to \infty$, where B is a standard Brownian motion, $L^a(0)$ (resp. $L^a(a^-)$) is the local time of H^a at level 0 (resp. at level a), and

$$H_s^a = \frac{2}{\sigma^2}(\alpha - \beta)s + \frac{2}{\sigma}B_s + \frac{1}{2}\left[L_s^a(0) - L_s^a(a^-)\right],$$

i.e., H^a *equals* $2/\sigma$ *multiplied by Brownian motion with drift* $(\alpha - \beta)s/\sigma$, *reflected in the interval* $[0, a]$.

5.4 A Ray–Knight Theorem

In this section we give a new proof of Delmas' generalization of the second Ray–Knight Theorem, see [16]. In case $\alpha \leq \beta$, we can choose $a = +\infty$, let $L_{\cdot}(0)$ denote the local time of H at level 0, and define

$$\tau_x = \inf\left\{s > 0, \, L_s(0) > \frac{4}{\sigma^2}x\right\}.$$

In the supercritical case, of course the construction is more complex. It follows from Lemma 11 and Corollary 3 (see also Lemma 2.1 in [16]) that for any $0 < a < b$,

$$\Pi^{a,b}(H^b) \stackrel{(d)}{=} H^a, \tag{5.14}$$

where H^a [resp. H^b] is $2/\sigma$ multiplied by Brownian motion, with drift $(\alpha - \beta)s/\sigma$, reflected in the interval $[0, a]$ [resp. $[0, b]$], see Theorem 7. Now define for each $a, x > 0$,

$$\tau_x^a = \inf\left\{s > 0, \, L_s^a(0) > \frac{4}{\sigma^2}x\right\}.$$

It follows from (5.14) that, as in the discrete case, $\forall\, 0 < a < b$,

$$\{L_{\tau_x^b}^b(t), \, 0 \leq t < a, x > 0\} \stackrel{(d)}{=} \{L_{\tau_x^a}^a(t), \, 0 \leq t < a, x > 0\}.$$

Consequently we can define the projective limit, which is a process $\{\mathscr{L}_x(t), \, t \geq 0, x > 0\}$ such that for each $a > 0$,

$$\{\mathscr{L}_x(t), \, 0 \leq t < a, x > 0\} \stackrel{(d)}{=} \{L_{\tau_x^a}^a(t), \, 0 \leq t < a, x > 0\}.$$

We have the (see also Theorem 3.1 in Delmas [16])

Theorem 8 (Generalized Ray–Knight theorem).

$$\{\mathscr{L}_x(t), t \geq 0, x > 0\} \stackrel{(d)}{=} \left\{\frac{4}{\sigma^2}Z_t^x, t \geq 0, x > 0\right\},$$

where Z^x *is the Feller branching diffusion process, solution of the SDE*

$$Z_t^x = x + (\alpha - \beta)\int_0^t Z_r^x dr + \sigma \int_0^t \sqrt{Z_r^x}dB_r, \, t \geq 0.$$

PROOF: Since both sides have stationary independent increments in x, it suffices to show that for any $x > 0$,

$$\{\mathcal{L}_x(t),\, t \geq 0\} \overset{(d)}{=} \left\{\frac{4}{\sigma^2} Z_t^x,\, t \geq 0\right\}.$$

Fix an arbitrary $a > 0$. By applying the elementary "occupation times formula" to $H^{a,N}$ (which differs from (5.3) since (5.8) differs from (5.2)), and Lemma 12, we have for any $g \in C(\mathbb{R}_+)$ with support in $[0,a]$,

$$\frac{4}{\sigma^2} \int_0^{\tau_x^{a,N}} g(H_r^{a,N}) dr = \int_0^\infty g(t) L_{\tau_x^{a,N}}^{a,N}(t) dt$$

$$\overset{(d)}{=} \frac{4}{\sigma^2} \int_0^\infty g(t) Z_t^{N,x} dt \qquad (5.15)$$

We deduce clearly from Proposition 7 that

$$\int_0^\infty g(t) Z_t^{N,x} dt \Longrightarrow \int_0^\infty g(t) Z_t^x dt. \qquad (5.16)$$

Let us admit for a moment that as $N \to \infty$

$$\int_0^{\tau_x^{a,N}} g(H_r^{a,N}) dr \Longrightarrow \int_0^{\tau_x^a} g(H_r^a) dr, \qquad (5.17)$$

where $\tau_x^a = \inf\{s > 0,\, L_s^a(0) > x\}$.

From the occupation times formula for the continuous semimartingale $(H_s^a, s \geq 0)$ (see Proposition 34 below), we have that

$$\frac{4}{\sigma^2} \int_0^{\tau_x^a} g(H_r^a) dr = \int_0^\infty g(t) L_{\tau_x^a}^a(t) dt. \qquad (5.18)$$

We deduce from (5.15), (5.16), (5.17), and (5.18) that for any $g \in C(\mathbb{R}_+)$ with compact support in $[0,a]$,

$$\frac{4}{\sigma^2} \int_0^\infty g(t) Z_t^x dt \overset{(d)}{=} \int_0^\infty g(t) \mathcal{L}_x(t) dt.$$

Since both processes $(Z_t^x, t \geq 0)$ and $(\mathcal{L}_x(t), t \geq 0)$ are *a.s.* continuous, the theorem is proved. $\qquad \square$

It remains to prove (5.17), which clearly is a consequence of (recall the definition (5.9) of $\tau_x^{a,N}$)

Proposition 18.

$$(H^{a,N}, \tau_x^{a,N}) \Longrightarrow (H^a, \tau_x^a).$$

PROOF: For the sake of simplifying the notations, we suppress the superscript a. Let us define the function ϕ from $\mathbb{R}_+ \times C_\uparrow([0, +\infty))$ into $\mathbb{R}_+$ by

$$\phi(x, y) = \inf\{s > 0 : y(s) > x\}.$$

For any fixed x, the function $\phi(x, .)$ is continuous in the neighborhood of a function y which is strictly increasing at the time when it first reaches the value x. Define

$$\tau_x'^N := \phi\left(x, \frac{\sigma^2}{4} L_.^N(0)\right).$$

We note that for any $x > 0$, $s \longmapsto L_s(0)$ is a.s. strictly increasing at time τ_x, which is a stopping time. This follows from the strong Markov property, the fact that $H_{\tau_x} = 0$, and $L_\varepsilon(0) > 0$, for all $\varepsilon > 0$. Consequently τ_x is a.s. a continuous function of the trajectory $L_.(0)$, and from Corollary 3

$$(H^N, \tau_x'^N) \Longrightarrow (H, \tau_x).$$

It remains to prove that $\tau_x'^N - \tau_x^N \longrightarrow 0$ in probability. For any $y < x$, for N large enough

$$0 \le \tau_x'^N - \tau_x^N \le \tau_x'^N - \tau_y'^N.$$

Clearly $\tau_x'^N - \tau_y'^N \Longrightarrow \tau_x - \tau_y$, hence for any $\varepsilon > 0$,

$$0 \le \limsup_N \mathbb{P}\left(\tau_x'^N - \tau_x^N \ge \varepsilon\right) \le \limsup_N \mathbb{P}\left(\tau_x'^N - \tau_y'^N \ge \varepsilon\right) \le \mathbb{P}\left(\tau_x - \tau_y \ge \varepsilon\right).$$

The result follows, since $\tau_y \uparrow \tau_x$ a.s. as $y \uparrow x$, $y < x$. $\square$

Chapter 6
Models of Finite Population with Interaction

We now want to model the interaction between individuals in our population, so that the resulting process will no longer be a branching process. We consider in this chapter a continuous time model for a finite population with interaction, in which each individual, independently of the others, gives birth naturally at rate b and dies naturally at rate d. Moreover, we suppose that each individual gives birth and dies because of interaction with others at rates which depend upon the current population size. We exclude multiple births at any given time and we define the interaction rule through a continuous function f which again satisfies

Assumption (H1) $f(0) = 0$ and there exists a constant $\beta > 0$ such that for all $x, y > 0$, $f(x+y) - f(x) \le \beta y$.

In order to define our model jointly for all ancestral population sizes, we need to introduce an asymmetric description of the effect of the interaction. This is done in sections 6.1 and 6.2. In section 6.3, we describe the contour process of the associated genealogical tree. Section 6.4 answers the question whether or not, as an effect of the interaction, the extinction time and the length of the genealogical forest of trees remain finite in the limit of infinitely many ancestors at time 0. The last two sections prepare the next chapter, and describe the renormalization of the models of the evolution of the population size, and of the contour process of the genealogical forest of trees.

6.1 The Model

We consider a continuous time $\mathbb{Z}_+$-valued population process $\{X_t^m, t \ge 0\}$, which starts at time zero from m ancestors who are arranged from left to right, and evolves in continuous time. The left/right order is passed on to their offsprings: the daughters are placed on the right of their mothers and if at a time t the individual i is located at the left of individual j, then all the daughters of i after time t will be placed on

© Springer International Publishing Switzerland 2016
É. Pardoux, *Probabilistic Models of Population Evolution*, Mathematical
Biosciences Institute Lecture Series 1.6, DOI 10.1007/978-3-319-30328-4_6

the left of j and all of its daughters. Those rules apply inside each genealogical tree, and distinct branches of the tree never cross. Since we have excluded multiple births at any given time, this means that the forest of genealogical trees of the population is a plane forest of trees, where the ancestor of the population X_t^1 is placed on the far left, the ancestor of $X_t^2 - X_t^1$ immediately on her right, etc... This defines in a non-ambiguous way an order from left to right within the population alive at each time t. See Figure 6.1. We decree that each individual feels the interaction with the others placed on her left but not with those on her right. Precisely, at any time t, the individual i has an interaction death rate equal to $(f(\mathscr{L}_i(t)+1) - f(\mathscr{L}_i(t)))^-$ or an interaction birth rate equal to $(f(\mathscr{L}_i(t)+1) - f(\mathscr{L}_i(t)))^+$, where $\mathscr{L}_i(t)$ denotes the number of individuals alive at time t who are located on the left of i in the above planar picture. This means that the individual i is under attack by the others located at her left if $f(\mathscr{L}_i(t)+1) - f(\mathscr{L}_i(t)) < 0$ while the interaction improves her fertility if $f(\mathscr{L}_i(t)+1) - f(\mathscr{L}_i(t)) > 0$. Of course, conditionally upon $\mathscr{L}_i(\cdot)$, the occurrence of a "competition death event" or an "interaction birth event" for individual i is independent of the other birth/death events and of what happens to the other individuals. In order to simplify our formulas, we suppose moreover that the first individual in the left/right order has a birth rate equal to $b + f^+(1)$ and a death rate equal to $d + f^-(1)$.

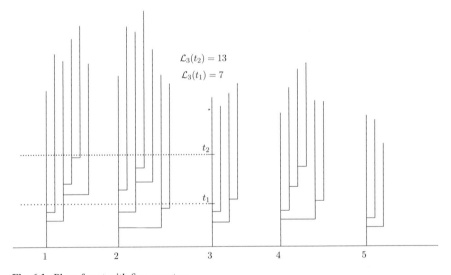

Fig. 6.1 Plane forest with five ancestors

The resulting total interaction birth rates minus the total interaction death rates endured by the population X_t^m at time t is then

$$\sum_{k=1}^{X_t^m} [(f(k) - f(k-1))^+ - (f(k) - f(k-1))^-] = \sum_{k=1}^{X_t^m} (f(k) - f(k-1)) = f(X_t^m).$$

As a result, $\{X_t^m,\, t \ge 0\}$ is a discrete-mass $\mathbb{Z}_+$-valued Markov process, which evolves as follows: $X_0^m = m$. If $X_t^m = 0$, then $X_s^m = 0$ for all $s \ge t$. While at state $k \ge 1$, the process

$$X_t^m \text{ jumps to } \begin{cases} k+1, & \text{at rate } bk + \sum_{\ell=1}^{k}(f(\ell)-f(\ell-1))^+; \\ k-1, & \text{at rate } dk + \sum_{\ell=1}^{k}(f(\ell)-f(\ell-1))^-. \end{cases}$$

6.2 Coupling over Ancestral Population Size

The above description specifies the joint evolution of all $\{X_t^m,\, t \ge 0\}_{m \ge 1}$, or in other words of the two-parameter process $\{X_t^m,\, t \ge 0, m \ge 1\}$. In the case of a linear function f, for each fixed $t > 0$, $\{X_t^m,\, m \ge 1\}$ is an independent increments process. We believe that there exist nonlinear functions f such that for t fixed $\{X_t^m,\, m \ge 1\}$ is not a Markov chain. That is to say, the conditional law of X_t^{n+1} given X_t^n differs from its conditional law given $(X_t^1, X_t^2, \ldots, X_t^n)$. The intuitive reason for that is that the additional information carried by $(X_t^1, X_t^2, \ldots, X_t^{n-1})$ gives us a clue as to the fertility or the level of competition that the progeny of the $n+1$st ancestor had to benefit or to suffer from, between time 0 and time t.

However, $\{X^m,\, m \ge 1\}$ is a Markov chain with values in the space $D([0,\infty);\mathbb{Z}_+)$ of càdlàg functions from $[0,\infty)$ into $\mathbb{Z}_+$, which starts from 0 at $m = 0$. Consequently, in order to describe the law of the whole process, that is, of the two-parameter process $\{X_t^m,\, t \ge 0, m \ge 1\}$, it suffices to describe the conditional law of X^n, given X^{n-1} for each $n \ge 1$. We now describe the conditional law of X^n given X^m for arbitrary $1 \le m < n$. Let $V_t^{m,n} := X_t^n - X_t^m, t \ge 0$. Conditionally upon $\{X^\ell,\, \ell \le m\}$, and given that $X_t^m = x(t), t \ge 0$, $\{V_t^{m,n},\, t \ge 0\}$ is a $\mathbb{Z}_+$-valued time inhomogeneous Markov process starting from $V_0^{m,n} = n - m$, whose time-dependent infinitesimal generator $\{Q_{k,\ell}(t),\, k, \ell \in \mathbb{Z}_+\}$ is such that its off-diagonal terms are given by

$$Q_{0,\ell}(t) = 0, \quad \forall \ell \ge 1, \quad \text{and for any } k \ge 1,$$

$$Q_{k,k+1}(t) = bk + \sum_{j=1}^{k}(f(x(t)+j)-f(x(t)+j-1))^+,$$

$$Q_{k,k-1}(t) = dk + \sum_{j=1}^{k}(f(x(t)+j)-f(x(t)+j-1))^-,$$

$$Q_{k,\ell}(t) = 0, \quad \forall \ell \notin \{k-1,k,k+1\}.$$

This description of the conditional law of $\{X_t^n - X_t^m,\, t \ge 0\}$, given X^m, is prescribed by what we have said above, and $\{X^m,\, m \ge 1\}$ is indeed a Markov chain.

Remark 10. Note that if the function f is increasing on $[0, a]$, $a > 0$, and decreasing on $[a, \infty)$, then the interaction improves the rate of fertility in a population whose size is smaller than a, but for large size the interaction amounts to competition within the population. This is reasonable since when the population is large, the limitation of resources implies competition within the population. A positive interaction (for moderate population sizes) has been discovered by Warder Clyde Allee in the 1930's who noticed that goldfish grow more rapidly when there are more individuals within a tank. Indeed, aggregation can improve the survival rate of individuals, and cooperation may be crucial in the overall evolution of social structure. This is called the Allee effect. We are mainly interested in the model with interaction defined with functions f such that $\lim_{x\to\infty} f(x) = -\infty$. Note also that we could have generalized our model to the case $f(0) \geq 0$. $f(0) > 0$ would mean an immigration flux. The reader can easily check that part of our results would still be valid in this case. However when we study extinction, the assumption $f(0) = 0$ is crucial.

6.3 The Associated Contour Process in the Discrete Model

The just described reproduction dynamics give rise to a forest $\mathscr{F}^m$ of m trees, drawn into the plane as sketched in Figure 6.2. Note also that, with the above described construction, the $(\mathscr{F}^m, m \geq 1)$ are coupled: the forest $\mathscr{F}^{m+1}$ has the same law as the forest $\mathscr{F}^m$ to which we add a new tree generated by an ancestor placed at the $(m+1)$st position. If the function f tends to $-\infty$ and m is large enough, the trees further to the right of the forest $\mathscr{F}^m$ have a tendency to stay smaller because of the competition : they are "under attack" from the trees to their left. From $\mathscr{F}^m$ we read off a continuous and piecewise linear $\mathbb{R}_+$-valued path $H^m = (H^m_s)$ (called the contour process of $\mathscr{F}^m$) which is described as follows.

Starting from 0 at the initial time $s = 0$, the process H^m rises at speed p until it hits the top of the first ancestor branch (this is the leaf marked with D in Figure 6.2). There it turns and goes downwards, now at speed $-p$, until arriving at the next branch point (which is B in Figure 6.2). From there it goes upwards into the (yet unexplored) next branch, and proceeds in a similar fashion until being back at height 0, which means that the contour of the leftmost tree is completed. Then explore the next tree, and so on. See Figure 6.2.

We define the local time $L^m_s(t)$ accumulated by the process H^m at level t up to time s by

$$L^m_s(t) = \lim_{\varepsilon \to 0} \frac{1}{\varepsilon} \int_0^s 1_{t \leq H^m_r < t+\varepsilon} dr.$$

The process H^m is piecewise linear, continuous with derivative $\pm p$: at any time $s \geq 0$, the rate of appearance of minima (giving rise to births, i.e., to the creation of new branches) is equal to

$$pb + p \left[f\left(\lfloor \tfrac{p}{2} L^m_s(H^m_s) \rfloor + 1 \right) - f\left(\lfloor \tfrac{p}{2} L^m_s(H^m_s) \rfloor \right) \right]^+, \tag{6.1}$$

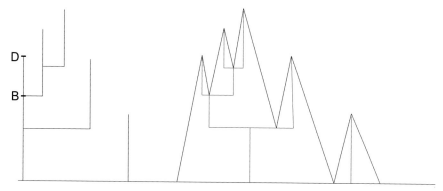

Fig. 6.2 A forest with two trees and its contour process.

and the rate of appearance of maxima (describing deaths of branches) is equal to

$$pd + p\left[f\left(\lfloor\frac{p}{2}L_s^m(H_s^m)\rfloor + 1\right) - f\left(\lfloor\frac{p}{2}L_s^m(H_s^m)\rfloor\right)\right]^-. \qquad (6.2)$$

Let τ^m be the time needed in order to explore the forest $\mathscr{F}^m$. We have

$$\tau^m = \inf\{s > 0; \frac{p}{2}L_s^m(0) \geq m\}.$$

Under the assumption that $\tau^m < \infty$ *a.s.* for all $m \geq 1$, if we choose above $p = 2$, we have the following discrete Ray–Knight representation (see Figure 6.3).

Corollary 4. *We have the following identity*

$$(X_t^m, t \geq 0, m \geq 1) \equiv (L_{\tau^m}^m(t), t \geq 0, m \geq 1).$$

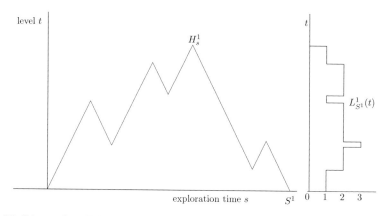

Fig. 6.3 Discrete Ray–Knight representation.

Remark 11. The condition that $\tau^m < \infty$ a.s. for all $m \geq 1$ is equivalent to the fact that the population X_t^m goes extinct in finite time a.s. for all ancestral population sizes $m \geq 1$. If this condition is not met, we can reflect the contour process below a given level, as we did above in the case without interaction in the supercritical case. This will be done below in section 6.6.

Remark 12. We note that the process $\{H_s^m,\ s \geq 0\}$ is not a Markov process. Its evolution after time s does not depend only upon the present value H_s^m, but also upon its past before s, through the local times accumulated up to time s.

6.4 The Effect of Interactions for Large Population

Define the lifetime of the population (which is also the height of the forest of genealogical trees), and the length of the forest of genealogical trees by

$$T^m = \inf\{t > 0, X_t^m = 0\}, \qquad L^m = \int_0^{T^m} X_t^m dt, \qquad \text{for} \quad m \geq 1.$$

Clearly T^m and L^m are a.s. increasing functions of m. We now study the limits of T^m and L^m as $m \to \infty$. We first recall some preliminary results on birth and death processes, which can be found in [3, 14, 22].

Let Y be a birth and death process with birth rate $\lambda_n > 0$ and death rate $\mu_n > 0$ when in state $n, n \geq 1$. Let

$$A = \sum_{n \geq 1} \frac{1}{\pi_n}, \qquad S = \sum_{n \geq 1} \frac{1}{\pi_n} \sum_{k \geq n+1} \frac{\pi_k}{\lambda_k},$$

where

$$\pi_1 = 1, \qquad \pi_n = \frac{\lambda_1 \ldots \lambda_{n-1} \lambda_n}{\mu_2 \ldots \mu_n}, \quad n \geq 2.$$

We denote S_y^m the first time the process Y hits $y \in [0, \infty)$ when starting from $Y_0 = m$.

$$S_y^m = \inf\{t > 0 : Y_t = y \mid Y_0 = m\}.$$

We say that ∞ is an entrance boundary for Y (see, for instance, Anderson [3], section 8.1) if there is $y > 0$ and a time $t > 0$ such that

$$\lim_{m \uparrow \infty} \mathbb{P}(S_y^m < t) > 0.$$

We have the following result (see [14], Proposition 7.10)

Proposition 19. *The following are equivalent:*

1) ∞ *is an entrance boundary for* Y.
2) $A = \infty, S < \infty$.
3) $\lim_{m \uparrow \infty} \mathbb{E}(S_0^m) < \infty$.

Now, we shall apply the above result to the process X_t^m, in which case $\lambda_n = bn + F^+(n), \mu_n = dn + F^-(n), n \geq 1$, where $F^+(n) = \sum_{\ell=1}^n (f(\ell) - f(\ell-1))^+$ and $F^-(n) = \sum_{\ell=1}^n (f(\ell) - f(\ell-1))^-$. We will need the following Lemmas.

Lemma 16. *Let f be a continuous function satisfying (H1), $a \in \mathbb{R}$ be a constant. If there exists $a_0 > 0$ such that $f(x) \neq 0, f(x) + ax \neq 0$ for all $x \geq a_0$, we have*

$$\int_{a_0}^\infty \frac{1}{|f(x)|} dx < \infty \Leftrightarrow \int_{a_0}^\infty \frac{1}{|ax + f(x)|} dx < \infty,$$

and when those equivalent conditions are satisfied, we have

$$\lim_{x\to\infty} \frac{f(x)}{x} = -\infty.$$

PROOF: We need only show that

$$\int_{a_0}^\infty \frac{1}{|f(x)|} dx < \infty \Rightarrow \int_{a_0}^\infty \frac{1}{|ax + f(x)|} dx < \infty.$$

Indeed, this will imply the same implication for pair $f'(x) = f(x) + ax$ and $f'(x) - ax$, which is the converse. Because $f(x) \leq \beta x$ for all $x \geq 0$, we can easily deduce from $\int_{a_0}^\infty \frac{1}{|f(x)|} dx < \infty$ that

$$f(x) < 0 \quad \text{for all } x \geq a_0.$$

Let β' be a constant such that $\beta' > \beta$. We have

$$\int_{a_0}^\infty \frac{1}{\beta' x - f(x)} dx < \int_{a_0}^\infty \frac{1}{-f(x)} dx < \infty.$$

This implies that

$$\lim_{x\to\infty} \int_x^{2x} \frac{1}{\beta' u - f(u)} du = 0.$$

Note that the function $\beta' x - f(x)$ is increasing, we deduce that $\lim_{x\to\infty} \frac{f(x)}{x} = -\infty$. Hence there exists $a_1 > a_0$ such that $f(x) < -2|a|x$ for all $x \geq a_1$. The result follows from

$$\int_{a_1}^\infty \frac{1}{|ax + f(x)|} dx < \int_{a_1}^\infty \frac{2}{-f(x)} dx < \infty.$$

$\square$

Lemma 17. *Let f be a function satisfying (H1). For all $n \geq 1$ we have the two inequalities*

$$F^+(n) \leq \beta n$$
$$-f(n) \leq F^-(n) \leq \beta n - f(n).$$

PROOF: The result follows from the facts that for all $n \geq 1$

$$(f(n) - f(n-1))^+ \leq \beta$$
$$(f(n) - f(n-1))^- \geq f(n-1) - f(n)$$
$$F^-(n) - F^+(n) = -f(n).$$

$\square$

Proposition 20. *Assume that f is a continuous function satisfying (H1) and there exists $a_0 > 0$ such that $f(x) \neq 0$ for all $x \geq a_0$. Then ∞ is an entrance boundary for X if and only if*

$$\int_{a_0}^{\infty} \frac{1}{|f(x)|} dx < \infty.$$

PROOF: If $\int_{a_0}^{\infty} \frac{1}{|f(x)|} dx = \infty$, then (recall that $(d+\beta)x - f(x)$ is nonnegative and increasing) $\int_{a_0}^{\infty} \frac{1}{(d+\beta)x - f(x)} dx = \infty$, by Lemma 16. In this case,

$$S \geq \sum_{n \geq 1} \frac{\pi_{n+1}}{\lambda_{n+1} \pi_n}$$

$$= \sum_{n \geq 1} \frac{1}{\mu_{n+1}}$$

$$= \sum_{n \geq 2} \frac{1}{dn + F^-(n)}$$

$$\geq \sum_{n \geq 2} \frac{1}{(d+\beta)n - f(n)}$$

$$= \infty.$$

Therefore, ∞ is not an entrance boundary for X, by Proposition 19. On the other hand, if $\int_{a_0}^{\infty} \frac{1}{|f(x)|} dx < \infty$, then $\lim_{x \to \infty} \frac{f(x)}{x} = -\infty$, by Lemma 16. By Lemma 17 we have

$$\lim_{n \to \infty} \frac{\pi_{n+1}}{\pi_n} = \lim_{n \to \infty} \frac{bn + F^+(n)}{dn + F^-(n)} \leq \lim_{n \to \infty} \frac{(b+\beta)n}{dn - f(n)} = 0,$$

so that

$$A = \sum_{n \geq 1} \frac{1}{\pi_n} = \infty.$$

Set $a_n = \lambda_n / \mu_n$, then there exists $n_0 \geq 1$ such that $a_n < 1$ for all $n \geq n_0$. The inequality of arithmetic and geometric means states that for all $m > 0$ and $x_1, x_2, \ldots, x_m > 0$,

$$\frac{x_1 + x_2 + \ldots + x_m}{m} \geq \sqrt[m]{x_1 x_2 \ldots x_m},$$

so that for all $k > n > 0$,

$$a_{n+1}^{k-n} + \ldots + a_k^{k-n} \geq (k-n) a_{n+1} \ldots a_k.$$

Then

$$
\begin{aligned}
\sum_{n\geq n_0} \frac{1}{\pi_n} \sum_{k\geq n+1} \frac{\pi_k}{\lambda_k} &\leq \frac{1}{b} \sum_{n\geq n_0} \sum_{k\geq n+1} \frac{1}{k} a_{n+1}\ldots a_k \\
&\leq \frac{1}{b} \sum_{n\geq n_0} \sum_{k\geq n+1} \frac{1}{k(k-n)} (a_{n+1}^{k-n} + \ldots + a_k^{k-n}) \\
&= \frac{1}{b} \sum_{k\geq n_0+1} \sum_{n=1}^{k-n_0} \frac{1}{kn} (a_{k-n+1}^n + \ldots + a_k^n) \\
&= \frac{1}{b} \sum_{i\geq n_0+1} \sum_{n\geq 1} a_i^n \sum_{k=i}^{n-1+i} \frac{1}{kn} \\
&\leq \frac{1}{b} \sum_{i\geq n_0+1} \sum_{n\geq 1} \frac{a_i^n}{i} \\
&= \frac{1}{b} \sum_{i\geq n_0+1} \frac{a_i}{i(1-a_i)} \\
&= \frac{1}{b} \sum_{i\geq n_0+1} \frac{\lambda_i}{i(\mu_i - \lambda_i)} \\
&= \sum_{i\geq n_0+1} \frac{bi + F^+(i)}{bi(di - bi + F^-(i) - F^+(i))} \\
&\leq \frac{b+\beta}{b} \sum_{i\geq n_0+1} \frac{1}{(d-b)i - f(i)} \\
&< \infty,
\end{aligned}
$$

where we have used Lemma 16 to conclude. Hence $S < \infty$. The result follows from Proposition 19. $\qquad\square$

The combination of Proposition 19 and Proposition 20 yields

Theorem 9. *Assume that f is a continuous function satisfying (H1) and there exists $a_0 > 0$ such that $f(x) \neq 0$ for all $x \geq a_0$. We have*

1) *If $\int_{a_0}^{\infty} \frac{1}{|f(x)|} dx = \infty$, then*

$$
\sup_{m>0} S_0^m = \infty \quad a.s.
$$

2) *If $\int_{a_0}^{\infty} \frac{1}{|f(x)|} dx < \infty$, then*

$$
\mathbb{E}\left(\sup_{m>0} S_0^m \right) < \infty.
$$

Remark 13. The first part of Theorem 9 is still true when $\lambda_n = 0, n \geq 1$. In fact, in this case we have

$$
S_0^m \overset{(d)}{=} \sum_{n=1}^{m} \theta_n,
$$

where $\overset{(d)}{=}$ denotes equality in law, θ_n represents the first passage time from state n to state $n-1$,

$$\theta_n = \inf\{t > 0 : X_t = n-1 \mid X_0 = n\}.$$

Recalling the fact that θ_n is exponentially distributed with parameter $\mu n + F^-(n)$, we have (see Lemma 4.3, Chapter 7 in [34])

$$\sup_{m>0} S_0^m = \infty \quad a.s. \quad \Leftrightarrow \quad \sum_{n=1}^{\infty} \frac{1}{\mu n + F^-(n)} = \infty.$$

The result follows by Lemma 16 and Lemma 17.

We now improve our result under the condition $\int_{a_0}^{\infty} \frac{1}{|f(x)|} dx < \infty$.

Theorem 10. *Suppose that f is a continuous function satisfying (H1) and there exists $a_0 > 0$ such that $f(x) \neq 0$ for all $x \geq a_0$. If $\int_{a_0}^{\infty} \frac{1}{|f(x)|} dx < \infty$, we have*

1) *For any $a > 0$, there exists $y_a \in \mathbb{Z}_+$ such that*

$$\sup_{m>y_a} \mathbb{E}\left(e^{aS_{y_a}^m}\right) < \infty.$$

2) *There exists some positive constant c such that*

$$\sup_{m>0} \mathbb{E}\left(e^{cS_0^m}\right) < \infty.$$

PROOF:

1) There exists $n_a \in \mathbb{Z}_+$ large enough so that

$$\sum_{n=n_a}^{\infty} \frac{1}{\pi_n} \sum_{k \geq n+1} \frac{\pi_k}{\lambda_k} \leq \frac{1}{a}.$$

Let J be the nonnegative increasing function defined by

$$J(m) := \sum_{n=n_a}^{m-1} \frac{1}{\pi_n} \sum_{k \geq n+1} \frac{\pi_k}{\lambda_k}, \qquad m \geq n_a + 1.$$

Set now $y_a = n_a + 1$. Note that $\sup_{m>y_a} S_{y_a}^m < \infty$ a.s., then for any $m > y_a$ we have

$$J(X_{t \wedge S_{y_a}^m}^m) - J(m) - \int_0^{t \wedge S_{y_a}^m} AJ(X_s^m) ds$$

is a martingale, where A is the generator of the process X_t^m which is given by

$$Ag(n) = \lambda_n(g(n+1) - g(n)) + \mu_n(g(n-1) - g(n)), \qquad n \geq 1,$$

for any $\mathbb{R}_+$-valued, bounded function g. It follows that

$$e^{a(t \wedge S^m_{y_a})} J(X^m_{t \wedge S^m_{y_a}}) - J(m) - \int_0^{t \wedge S^m_{y_a}} e^{as}(aJ(X^m_s) + AJ(X^m_s))ds$$

is also a martingale. It implies that

$$\mathbb{E}\left(e^{a(t \wedge S^m_{y_a})} J(X^m_{t \wedge S^m_{y_a}})\right) = J(m) + \mathbb{E}\left(\int_0^{t \wedge S^m_{y_a}} e^{as}(aJ(X^m_s) + AJ(X^m_s))ds\right).$$

We have for $m > y_a$, $J(X^m_s) < J(\infty) \leq \frac{1}{a} \quad \forall s \leq S^m_{y_a}$, and for any $n \geq y_a$,

$$AJ(n) = \lambda_n(J(n+1) - J(n)) + \mu_n(J(n-1) - J(n))$$

$$= \lambda_n \frac{1}{\pi_n} \sum_{k \geq n+1} \frac{\pi_k}{\lambda_k} - \mu_n \frac{1}{\pi_{n-1}} \sum_{k \geq n} \frac{\pi_k}{\lambda_k}$$

$$= \frac{\mu_2 \dots \mu_n}{\lambda_1 \dots \lambda_{n-1}} \sum_{k \geq n+1} \frac{\pi_k}{\lambda_k} - \frac{\mu_2 \dots \mu_n}{\lambda_1 \dots \lambda_{n-1}} \sum_{k \geq n} \frac{\pi_k}{\lambda_k}$$

$$= -\frac{\mu_2 \dots \mu_n}{\lambda_1 \dots \lambda_{n-1}} \frac{\pi_n}{\lambda_n}$$

$$= -1.$$

So that

$$\mathbb{E}\left(e^{a(t \wedge S^m_{y_a})} J(X^m_{t \wedge S^m_{y_a}})\right) \leq J(m) \qquad \forall m > y_a.$$

But J is increasing, then for any $m > y_a$ one gets

$$0 < J(y_a) \leq J(m) < J(\infty) \leq \frac{1}{a}.$$

From this we deduce that

$$\mathbb{E}\left(e^{a(t \wedge S^m_{y_a})}\right) \leq \frac{1}{aJ(y_a)} \qquad \forall m > y_a.$$

Hence

$$\mathbb{E}\left(e^{aS^m_{y_a}}\right) \leq \frac{1}{aJ(y_a)} \qquad \forall m > y_a,$$

by the monotone convergence theorem. The result follows.

2) Using the first result of the theorem, there exists a constant $M \in \mathbb{Z}_+$ such that

$$\sup_{m > M} \mathbb{E}\left(e^{S^m_M}\right) < \infty,$$

or $\mathbb{E}\left(e^{S_M}\right) < \infty$, where $S_M := \sup_{m > M} S^m_M$.

Given any fixed $T > 0$, let p denote the probability that starting from M at time $t = 0, X$ hits zero before time T. Clearly $p > 0$. Let ζ be a geometric random variable with success probability p, which is defined as follows. Let X start from M at time 0. If X hits zero before time T, then $\zeta = 1$. If not, we look the position X_T of X at time T.

If $X_T > M$, we wait until X goes back to M. The time needed is stochastically dominated by the random variable S_M, which is the time needed for X to descend to M, when starting from ∞. If however $X_T \leq M$, we start afresh from there, since the probability to reach zero in less than T is greater than or equal to p, for all starting points in the interval $(0, M]$.

So either at time T or at time less than $T + S_M$, we start again from a level which is less than or equal to M. If zero is reached during the next time interval of length T, then $\zeta = 2\dots$ Repeating this procedure, we see that $\sup_{m>0} S_0^m$ is stochastically dominated by

$$\zeta T + \sum_{i=0}^{\zeta} \eta_i,$$

where the random variables η_i are i.i.d., with the same law as S_M, globally independent of ζ. We have

$$\sup_{m>0} \mathbb{E}\left(e^{cS_0^m}\right) \leq \mathbb{E}\left(e^{c(\zeta T + \sum_{i=0}^{\zeta} \eta_i)}\right)$$

$$\leq \sqrt{\mathbb{E}\left(e^{2c\zeta T}\right)}\sqrt{\mathbb{E}\left(e^{2c\sum_{i=0}^{\zeta}\eta_i}\right)}.$$

Since ζ is a geometric (p) random variable, then

$$\mathbb{E}\left(e^{2c\zeta T}\right) = \frac{p}{1-p} \sum_{k=1}^{\infty} \left(e^{2cT}(1-p)\right)^k < \infty,$$

provided that $c < -\log(1-p)/2T$. Moreover

$$\mathbb{E}\left(e^{2c\sum_{i=0}^{\zeta}\eta_i}\right) = \sum_{k=1}^{\infty} \mathbb{E}\left(e^{2c\sum_{i=0}^{k}\eta_i}\right)\mathbb{P}(\zeta = k)$$

$$= \sum_{k=1}^{\infty} \left[\mathbb{E}\left(e^{2cS_M}\right)\right]^{k+1}\mathbb{P}(\zeta = k)$$

$$= \frac{p}{(1-p)^2} \sum_{k=1}^{\infty} \left[\mathbb{E}\left(e^{2cS_M}\right)(1-p)\right]^{k+1}.$$

Since $\mathbb{E}\left(e^{S_M}\right) < \infty$, it follows from the monotone convergence theorem that $\mathbb{E}\left(e^{2cS_M}\right) \to 1$ as $c \to 0$. Hence we can choose $0 < c < -\log(1-p)/2T$ such that

$$\mathbb{E}\left(e^{2cS_M}\right)(1-p) < 1,$$

in which case $\mathbb{E}\left(e^{2c\sum_{i=0}^{\zeta}\eta_i}\right) < \infty$. Then $\sup_{m>0} \mathbb{E}\left(e^{cS_0^m}\right) < \infty$. The result follows.

$\square$

The following result follows from Theorem 9 and Theorem 10

Theorem 11. *Suppose that f is a continuous function satisfying (H1) and there exists $a_0 > 0$ such that $f(x) \neq 0$ for all $x \geq a_0$. We have*

1) *If $\int_{a_0}^{\infty} \frac{1}{|f(x)|} dx = \infty$, then*

$$\sup_{m>0} T^m = \infty \quad a.s.$$

2) *If $\int_{a_0}^{\infty} \frac{1}{|f(x)|} dx < \infty$, then*

$$\sup_{m>0} T^m < \infty \quad a.s.,$$

and, moreover, there exists some positive constant c such that

$$\sup_{m>0} \mathbb{E}\left(e^{cT^m}\right) < \infty.$$

Concerning the length of the forest of genealogical trees we have

Theorem 12. *Suppose that the continuous function $\frac{f(x)}{x}$ satisfies (H1) and there exists $a_0 > 0$ such that $f(x) \neq 0$ for all $x \geq a_0$. We have*

1) *If $\int_{a_0}^{\infty} \frac{x}{|f(x)|} dx = \infty$, then*

$$\sup_{m>0} L^m = \infty \quad a.s.$$

2) *If $\int_{a_0}^{\infty} \frac{x}{|f(x)|} dx < \infty$, then*

$$\sup_{m>0} L^m < \infty \quad a.s.,$$

and, moreover, there exists some positive constant c such that

$$\sup_{m>0} \mathbb{E}\left(e^{cL^m}\right) < \infty.$$

PROOF: We define

$$A_t^m := \int_0^t X_r^m dr, \qquad \eta_t^m = \inf\{s > 0, A_s^m > t\},$$

and consider the process $U^m := X^m \circ \eta^m$. Let S^m be the stopping time defined by

$$S^m = \inf\{r > 0, U_r^m = 0\}.$$

Note that $S^m = L^m$, the length of the genealogical tree of the population X^m, since $S^m = \int_0^{T^m} X_r^m dr$. The process X^m can be expressed using two mutually independent standard Poisson processes P_1 and P_2, as

$$X_t^m = m + P_1\left(\int_0^t [\lambda X_r^m + F^+(X_r^m)] dr\right) - P_2\left(\int_0^t [\mu X_r^m + F^-(X_r^m)] dr\right).$$

Consequently the process $U^m := X^m \circ \eta^m$ satisfies

$$U_t^m = m + P_1\left(\int_0^t [\lambda + \frac{F^+(U_r^m)}{U_r^m}]dr\right) - P_2\left(\int_0^t [\mu + \frac{F^-(U_r^m)}{U_r^m}]dr\right).$$

On the interval $[0, S^m)$, $U_r^m \geq 1$, hence by Lemma 18 below we have

$$\mu + \frac{F^-(U_r^m)}{U_r^m} \leq (\mu + 2\beta)U_r^m - \frac{f(U_r^m)}{U_r^m} \leq (\mu + 2\beta)U_r^m + F_1^-(U_r^m),$$

where

$$f_1(n) := \frac{f(n)}{n}, \qquad F_1^-(n) := \sum_{k=1}^n (f_1(k) - f_1(k-1))^-, \qquad n \geq 1.$$

Then

$$U_t^m \geq m - P_2\left(\int_0^t [(\mu + 2\beta)U_r^m + F_1^-(U_r^m)]dr\right).$$

The first part of the theorem is now a consequence of Theorem 9 and Remark 13. For the second part of the theorem, we note that in the case $\int_{a_0}^\infty \frac{x}{|f(x)|}dx < \infty$, we have $\frac{f(x)}{x^2} \to -\infty$ as $x \to \infty$, by Lemma 16. Then there exists a constant $u > 0$ such that for all $n \geq u$,

$$\mu + \frac{F^-(n)}{n} \geq \frac{-f(n)}{n} \geq \beta n - \frac{f(n)}{2n}.$$

We can choose $\varepsilon \in (0,1)$ such that for all $1 \leq n \leq u$

$$\mu \geq \varepsilon(\beta n - \frac{f(n)}{2n}).$$

It implies that for all $n \geq 1$,

$$\mu + \frac{F^-(n)}{n} \geq \varepsilon(\beta n - \frac{f(n)}{2n}).$$

Let $f_2(x) := \frac{\varepsilon}{2}(\frac{f(x)}{x} - \beta x)$. Then f_2 is a negative and decreasing function, so that

$$F_2^+(n) := \sum_{k=1}^n (f_2(k) - f_2(k-1))^+ = 0,$$
$$F_2^-(n) := \sum_{k=1}^n (f_2(k) - f_2(k-1))^- = -f_2(n), \quad \forall n \geq 1$$

Hence by Lemma 18 we have on $[0, S^m)$,

$$U_t \leq m + P_1\left(\int_0^t [(\lambda + 2\beta)U_r^m + F_2^+(U_r^m)]dr\right) - P_2\left(\int_0^t [\frac{\varepsilon\beta}{2}U_r^m + F_2^-(U_r^m)]dr\right).$$

The result follows from Theorem 10.   □

It remains to prove

Lemma 18. *Suppose that the function $\frac{f(x)}{x}$ satisfies (H1). For all $n \geq 1$ we have the following inequalities*

$$F^+(n) \leq 2\beta n^2,$$
$$-f(n) \leq F^-(n) \leq 2\beta n^2 - f(n).$$

PROOF: Note that for all $k \geq 1$,

$$(f(k) - f(k-1))^+ = \left((k-1)(\frac{f(k)}{k} - \frac{f(k-1)}{k-1}) + \frac{f(k)}{k}\right)^+$$

$$\leq (k-1)\left(\frac{f(k)}{k} - \frac{f(k-1)}{k-1}\right)^+ + \left(\frac{f(k)}{k}\right)^+$$

$$\leq 2\beta k.$$

Then

$$F^+(n) \leq \sum_{k=1}^{n} 2\beta k = \beta n(n+1) \leq 2\beta n^2.$$

The result follows from the fact that for all $n \geq 1$

$$(f(n) - f(n-1))^- \geq f(n-1) - f(n)$$
$$F^-(n) - F^+(n) = -f(n).$$

$\square$

Example 1. Let $f(z) = -ae^{bx}$, with $a, b > 0$. If $b \leq 1$, then $T^m \to \infty$ and $L^m \to \infty$, as $m \to \infty$. If $1 < b \leq 2$, then there exists $c > 0$ such that $\sup_{m \geq 1} \mathbb{E}\left[e^{cT^m}\right] < \infty$, while $L^m \to \infty$ as $m \to \infty$. Finally if $b > 2$, then for some $c > 0$, both $\sup_{m \geq 1} \mathbb{E}\left[e^{cT^m}\right] < \infty$ and $\sup_{m \geq 1} \mathbb{E}\left[e^{cL^m}\right] < \infty$.

6.5 Renormalized Discrete Model

Now we proceed to a renormalization of this model. For $x \in \mathbb{R}_+$ and $N \in \mathbb{N}$, we choose $m = \lfloor Nx \rfloor$, $b = \frac{\sigma^2}{2}N$, and $d = \frac{\sigma^2}{2}N$, we multiply f by N and divide by N the argument of the function f. We attribute to each individual in the population a mass equal to $1/N$. Then the total mass process $Z^{N,x}$, which starts from $\frac{\lfloor Nx \rfloor}{N}$ at time $t = 0$, is a Markov process whose evolution can be described as follows.

$$Z^{N,x} \text{ jumps from } \frac{k}{N} \text{ to } \begin{cases} \frac{k+1}{N} \text{ at rate } \sigma^2 Nk/2 + N\sum_{i=1}^{k}\left(f(\frac{i}{N}) - f(\frac{i-1}{N})\right)^+ \\ \frac{k-1}{N} \text{ at rate } \sigma^2 Nk/2 + N\sum_{i=1}^{k}\left(f(\frac{i}{N}) - f(\frac{i-1}{N})\right)^-, \end{cases}$$

Clearly there exist two mutually independent standard Poisson processes P_1 and P_2 such that

$$Z_t^{N,x} = \frac{\lfloor Nx \rfloor}{N} + \frac{1}{N} P_1 \left(\int_0^t \left\{ \frac{\sigma^2}{2} N^2 Z_r^{N,x} + N \sum_{i=1}^{NZ_r^{N,x}} (f(\frac{i}{N}) - f(\frac{i-1}{N}))^+ \right\} dr \right)$$
$$- \frac{1}{N} P_2 \left(\int_0^t \left\{ \frac{\sigma^2}{2} N^2 Z_r^{N,x} + N \sum_{i=1}^{NZ_r^{N,x}} (f(\frac{i}{N}) - f(\frac{i-1}{N}))^- \right\} dr \right).$$

Consequently there exists a local martingale $M^{N,x}$ such that

$$Z_t^{N,x} = \frac{\lfloor Nx \rfloor}{N} + \int_0^t f(Z_r^{N,x}) dr + M_t^{N,x}. \tag{6.3}$$

Since $M^{N,x}$ is a purely discontinuous local martingale, its quadratic variation $[M^{N,x}]$ is given by the sum of the squares of its jumps, i.e.,

$$[M^{N,x}]_t = \frac{1}{N^2} \left[P_1 \left(\int_0^t \left\{ \frac{\sigma^2}{2} N^2 Z_r^{N,x} + N \sum_{i=1}^{NZ_r^{N,x}} (f(\frac{i}{N}) - f(\frac{i-1}{N}))^+ \right\} dr \right) \right.$$
$$\left. + P_2 \left(\int_0^t \left\{ \frac{\sigma^2}{2} N^2 Z_r^{N,x} + N \sum_{i=1}^{NZ_r^{N,x}} (f(\frac{i}{N}) - f(\frac{i-1}{N}))^- \right\} dr \right) \right]. \tag{6.4}$$

We deduce from (6.4) that the predictable quadratic variation $\langle M^{N,x} \rangle$ of $M^{N,x}$ is given by

$$\langle M^{N,x} \rangle_t = \int_0^t \left\{ \sigma^2 Z_r^{N,x} + \frac{1}{N} ||f||_{N,0,Z_r^{N,x}} \right\} dr, \tag{6.5}$$

where for any $z = \frac{k}{N}$, $z' = \frac{k'}{N}$, and $k \in \mathbb{Z}_+$ such that $k \leq k'$,

$$||f||_{N,z,z'} = \sum_{i=k+1}^{k'} |f(\frac{i}{N}) - f(\frac{i-1}{N})|. \tag{6.6}$$

We now describe the law of the pair $(Z^{N,x}, Z^{N,y})$, for any $0 < x < y$. Consider the pair of process $(Z^{N,x}, V^{N,x,y})$, which starts from $\left(\frac{\lfloor Nx \rfloor}{N}, \frac{\lfloor Ny \rfloor - \lfloor Nx \rfloor}{N} \right)$ at time $t = 0$, and whose dynamic is described by: $(Z^{N,x}, V^{N,x,y})$ jumps

$$\text{from } (\frac{i}{N}, \frac{j}{N}) \text{ to } \begin{cases} (\frac{i+1}{N}, \frac{j}{N}) \text{ at rate } \sigma^2 N i/2 + N \sum_{k=1}^{i} (f(\frac{k}{N}) - f(\frac{k-1}{N})^+ \\ (\frac{i-1}{N}, \frac{j}{N}) \text{ at rate } \sigma^2 N i/2 + N \sum_{k=1}^{i} (f(\frac{k}{N}) - f(\frac{k-1}{N}))^- \\ (\frac{i}{N}, \frac{j+1}{N}) \text{ at rate } \sigma^2 N j/2 + N \sum_{k=1}^{j} (f(\frac{i+k}{N}) - f(\frac{i+k-1}{N}))^+ \\ (\frac{i}{N}, \frac{j-1}{N}) \text{ at rate } \sigma^2 N j/2 + N \sum_{k=1}^{j} (f(\frac{i+k}{N}) - f(\frac{i+k-1}{N}))^- \end{cases}.$$

The process $V^{N,x,y}$ can be expressed as follows

$$
\begin{aligned}
V_t^{N,x,y} &= \frac{\lfloor Ny \rfloor - \lfloor Nx \rfloor}{N} \\
&+ \frac{1}{N} P^1 \left(N \int_0^t \sum_{k=1}^{NV_s^{N,x,y}} \left(f(Z_s^{N,x} + \frac{k}{N}) - f(Z_s^{N,x} + \frac{k-1}{N}) \right)^+ ds \right) \\
&- \frac{1}{N} P^2 \left(N \int_0^t \sum_{k=1}^{NV_s^{N,x,y}} \left(f(Z_s^{N,x} + \frac{k}{N}) - f(Z_s^{N,x} + \frac{k-1}{N}) \right)^- ds \right) \\
&+ \frac{1}{N} P^3 \left(\frac{\sigma^2}{2} N^2 \int_0^t V_s^{N,x,y} ds \right) - \frac{1}{N} P^4 \left(\frac{\sigma^2}{2} N^2 \int_0^t V_s^{N,x,y} ds \right),
\end{aligned}
\tag{6.7}
$$

where P^1, P^2, P^3, and P^4 are mutually independent standard Poisson processes which are globally independent of $\{Z_\cdot^{N,x'}, x' \leq x\}$. Consequently

$$
\begin{aligned}
V_t^{N,x,y} &= \frac{\lfloor Ny \rfloor - \lfloor Nx \rfloor}{N} \\
&+ \int_0^t \left[f(Z_r^{N,x} + V_r^{N,x,y}) - f(Z_r^{N,x}) \right] dr + M_t^{N,x,y},
\end{aligned}
\tag{6.8}
$$

where $M^{N,x,y}$ is a local martingale whose predictable quadratic variation $\langle M^{N,x,y} \rangle$ is given by

$$
\langle M^{N,x,y} \rangle_t = \int_0^t \left\{ \sigma^2 V_r^{N,x,y} + \frac{1}{N} ||f||_{N, Z_r^{N,x}, V^{N,x,y} + Z_r^{N,x}} \right\} dr.
\tag{6.9}
$$

Since $Z^{N,x}$ and $V^{N,x,y}$ never jump at the same time,

$$
[M^{N,x}, M^{N,x,y}] = 0, \text{ hence } \langle M^{N,x}, M^{N,x,y} \rangle = 0,
\tag{6.10}
$$

which implies that the martingales $M^{N,x}$ and $M^{N,x,y}$ are orthogonal.

Consequently, $Z^{N,x} + V^{N,x,y}$ solves the SDE

$$
Z_t^{N,x} + V_t^{N,x,y} = \frac{\lfloor Ny \rfloor}{N} + \int_0^t f(Z_r^{N,x} + V_r^{N,x,y}) dr + \tilde{M}_t^{N,x,y}.
$$

where $\tilde{M}^{N,x,y}$ is a local martingale with $\langle \tilde{M}^{N,x,y} \rangle$ given by

$$
\langle \tilde{M}^{N,x,y} \rangle_t = \langle M^{N,x} \rangle_t + \langle M^{N,x,y} \rangle_t = \langle M^{N,x+y} \rangle_t, \quad \forall t \geq 0.
$$

We then deduce that for any $x, y \in \mathbb{R}_+$ such $x \leq y$,

$$
Z^{N,x} + V^{N,x,y} \stackrel{(d)}{=} Z^{N,y}.
$$

It follows from (6.7) that, conditionally upon $\{Z_\cdot^{N,x'}, x' \leq x\}$, $M^{N,x,y}$ is a local martingale.

6.6 Renormalization of the Contour Process

We now choose $b_N = d_N = \sigma^2 N/2$, and choose the slopes to be $\pm 2N$. Moreover we replace the function f by $f_N = N f(\cdot/N)$. Hence in the renormalized contour process H^N, the maxima appear at rate

$$
\sigma^2 N^2 + 2N^2 \left[f\left(\frac{\sigma^2}{4} L_s^N(H_s^N) + \frac{1}{N} \right) - f\left(\frac{\sigma^2}{4} L_s^N(H_s^N) \right) \right]^-,
$$

and the minima at rate

$$
\sigma^2 N^2 + 2N^2 \left[f\left(\frac{\sigma^2}{4} L_s^N(H_s^N) + \frac{1}{N} \right) - f\left(\frac{\sigma^2}{4} L_s^N(H_s^N) \right) \right]^+,
$$

where

$$
L_s^N(t) = \frac{4}{\sigma^2} \lim_{\varepsilon \to 0} \frac{1}{\varepsilon} \int_0^s \mathbf{1}_{\{t < H_r^N < t + \varepsilon\}} dr. \tag{6.11}
$$

Note that $\frac{\sigma^2}{4} L_s^N(t)$ equals the number of pairs of branches of the zigzag curve which hit the level t before time s, divided by N. The choice $b_N = d_N$ corresponds to a critical branching model in the absence of interactions.

We now want to use Girsanov's theorem, in order to reduce the present model to the one studied in section 5.3. As in that section,

$$
\frac{dH_s^N}{ds} = 2N V_s^N,
$$

where V_s^N takes values in $\{-1, 1\}$, and solves (with notations slightly different from those in the above section)

$$
dV_s^N = 2 dQ_s^{1,N} - 2 dQ_s^{2,N} + 2N dL_s^N(0),
$$

where

$$
Q_s^{1,N} = \int_0^s \mathbf{1}_{V_r^N = -1} dP_r^N, \quad Q_s^{2,N} = \int_0^s \mathbf{1}_{V_r^N = 1} dP_r^N,
$$

P_s^N being a Poisson point process with intensity $\sigma^2 N^2$ under the probability measure $\mathbb{P}$, so that $Q_s^{1,N}$ (resp., $Q_s^{2,N}$) has the intensity

$$
\lambda_r^{1,N} = \sigma^2 N^2 \mathbf{1}_{V_r^N = -1}, \quad \text{resp. } \lambda_r^{2,N} = \sigma^2 N^2 \mathbf{1}_{V_r^N = 1}.
$$

We define moreover the martingales

$$
M_r^N = P_r^N - \sigma^2 N^2 r, \tag{6.12}
$$

$$
\mathcal{M}_s^{1,N} = \frac{2}{N\sigma^2} \int_0^s \mathbf{1}_{V_r^{a,N} = -1} dM_r^N, \quad \mathcal{M}_s^{2,N} = \frac{2}{N\sigma^2} \int_0^s \mathbf{1}_{V_r^{a,N} = 1} dM_r^N,
$$

and the collection of σ-algebras $\mathcal{G}_s^N = \sigma\{H_r^N, 0 \le r \le s\}$.

We next introduce a Girsanov–Radon–Nikodym derivative.

$$Y_s^N = 1 + \int_0^s Y_{r-}^N \left[(f_N')^+ \left(\frac{\sigma^2}{4} L_{r-}^N (H_r^N) \right) d\mathcal{M}_r^{1,N} + (f_N')^- \left(\frac{\sigma^2}{4} L_{r-}^N (H_r^N) \right) d\mathcal{M}_r^{2,N} \right],$$

with $f_N'(x) = N[f(x+1/N) - f(x)]$. Under the additional assumption that f' is bounded, $\mathbb{E}Y_s^N = 1$ for all $s > 0$.

Define $\widetilde{\mathbb{P}}^N$ such that for each $s > 0$, if $\mathscr{G}_s^N = \sigma\{H_r^N, 0 \leq r \leq s\}$,

$$\frac{d\widetilde{\mathbb{P}}^N}{d\mathbb{P}} \bigg|_{\mathscr{G}_s} = Y_s^N.$$

It follows from Proposition 36 below with

$$\mu_r^{1,N} = 1 + \frac{2}{N\sigma^2} (f_N')^+ \left(\frac{\sigma^2}{4} L_{r-}^N (H_r^N) \right) \quad \text{and} \quad \mu_r^{2,N} = 1 + \frac{2}{N\sigma^2} (f_N')^- \left(\frac{\sigma^2}{4} L_{r-}^N (H_r^N) \right)$$

that under $\widetilde{\mathbb{P}}^{a,N}$, $Q_s^{1,N}$ (resp., $Q_s^{2,N}$) has the intensity

$$\left[\sigma^2 N^2 + 2N(f_N')^+ \left(\frac{\sigma^2}{4} L_{r-}^N (H_r^N) \right) \right] \mathbf{1}_{V_{r-}^N = -1},$$

$$\text{resp.} \quad \left[\sigma^2 N^2 + 2N(f_N')^- \left(\frac{\sigma^2}{4} L_{r-}^N (H_r^N) \right) \right] \mathbf{1}_{V_{r-}^N = 1}.$$

Note that if we decide to reflect the contour process below a given level $a > 0$, we just need to replace the above system of equations by

$$\frac{dH_s^{a,N}}{ds} = 2NV_s^{a,N},$$

$$dV_s^{a,N} = 2\mathbf{1}_{\{V_{s-}^{a,N} = -1\}} dP_s^N - 2\mathbf{1}_{\{V_{s-}^{a,N} = 1\}} dP_s^N + \frac{\sigma^2}{2} N dL_s^{a,N}(0) - \frac{\sigma^2}{2} N dL_s^{a,N}(a^-), \tag{6.13}$$

where $L_s^{a,N}(t)$ denotes the local time accumulated by the process $H^{a,N}$ at level t up to time s, defined by (6.11) with H^N replaced by $H^{a,N}$. The difference in the notations at levels 0 and a is a consequence of the fact that, as usual, we assume that $t \to L_s^{a,N}(t)$ is right-continuous. This, combined with the fact that the process $H^{a,N}$ lives in the interval $[0,a]$, implies that $L_s^{a,N}(0^-) = L_s^{a,N}(a) = 0$.

In this new framework we replace Y^N by $Y^{a,N}$, which is defined exactly as Y^N, but with (H^N, L^N, V^N) replaced by $(H^{a,N}, L^{a,N}, V^{a,N})$, that is,

$$Y_s^{a,N} = 1 + \int_0^s Y_{r-}^{a,N} \left[(f_N')^+ \left(\frac{\sigma^2}{4} L_{r-}^{a,N} (H_r^{a,N}) \right) d\mathcal{M}_r^{1,a,N} + (f_N')^- \left(\frac{\sigma^2}{4} L_{r-}^{a,N} (H_r^{a,N}) \right) d\mathcal{M}_r^{2,a,N} \right], \tag{6.14}$$

where

$$\mathscr{M}_s^{1,a,N} = \frac{2}{N\sigma^2} \int_0^s \mathbf{1}_{V_r^{a,N} = -1} dM_r^N, \quad \mathscr{M}_s^{2,a,N} = \frac{2}{N\sigma^2} \int_0^s \mathbf{1}_{V_r^{a,N} = 1} dM_r^N. \quad (6.15)$$

Under the additional assumption that f' is bounded, it is clear that $Y^{a,N}$ is a martingale, hence $\mathbb{E}[Y_s^{a,N}] = 1$ for all $s \geq 0$. In this case, we define $\widetilde{\mathbb{P}}^{a,N}$ as the probability such that for each $s < \infty$,

$$\frac{d\widetilde{\mathbb{P}}^{a,N}}{d\mathbb{P}}\Bigg|_{\mathscr{F}_s^{a,N}} = Y_s^{a,N},$$

where $\mathscr{F}_s^{a,N} := \sigma\{H_r^{a,N}, 0 \leq r \leq s\}$. Define

$$Q_s^{1,a,N} = \int_0^s \mathbf{1}_{V_r^{a,N} = -1} dP_r^N, \quad Q_s^{2,a,N} = \int_0^s \mathbf{1}_{V_r^{a,N} = 1} dP_r^N.$$

It follows from Theorem 36 below that under $\widetilde{\mathbb{P}}^{a,N}$, $Q_s^{1,a,N}$, resp., $Q_s^{2,a,N}$, has the intensity

$$\left[\sigma^2 N^2 + 2N(f_N')^+ \left(\frac{\sigma^2}{4} L_{r^-}^{a,N}(H_r^{a,N})\right)\right] \mathbf{1}_{V_{r^-}^{a,N} = -1},$$

$$\text{resp. } \left[\sigma^2 N^2 + 2N(f_N')^- \left(\frac{\sigma^2}{4} L_{r^-}^{a,N}(H_r^{a,N})\right)\right] \mathbf{1}_{V_{r^-}^{a,N} = 1}.$$

Let

$$\tau_x^{a,N} = \inf\left\{s > 0, L_s^{a,N}(0) > \frac{4}{\sigma^2}\lfloor Nx \rfloor / N\right\}.$$

Corollary 4 reads in this situation

Corollary 5. *The law of*

$$\left\{\frac{4}{\sigma^2} Z_t^{N,x}, 0 \leq t < a, x > 0\right\}$$

coincides with the law of

$$\{L_{\tau_x^{a,N}}^{a,N}(t), 0 \leq t < a, x > 0\}$$

under $\widetilde{\mathbb{P}}^{a,N}$.

Let now $0 < a < b$. Recall the definition of the mapping $\Pi^{a,b}$ in (5.5). We have the

Lemma 19. *For any* $0 < a < b, N \geq 1$,

$$\widetilde{\mathbb{P}}^{a,N} = \widetilde{\mathbb{P}}^{b,N}(\Pi^{a,b})^{-1}.$$

PROOF: The argument is essentially the same as that of Lemma 11, taking into account that the interaction rate while below a depends only upon the past visits of $H^{b,N}$ at the same level, and not upon the trajectory above level a. $\qquad\square$

Chapter 7
Convergence to a Continuous State Model

The aim of this chapter is to take the limit in the renormalized version of the model of the previous chapter, i.e., let $N \to \infty$ in the models of sections 6.5 and 6.6, respectively. In section 7.1, we shall take the limit in the model for the evolution of the population size, as a function of the two parameters x (the ancestral population size) and t (time). Since we want to stick to our rather minimal assumptions on the function f, checking tightness requires some care. In section 7.2, we will take the limit in the renormalized contour process of section 6.6. Here there are two difficulties. One is the fact that since we do not want to restrict ourself to the (sub)critical case, it is not clear whether the contour process will accumulate an arbitrary amount of local time at level 0. In order to circumvent this difficulty, we use as in chapter 5 a trick due to Delmas [16] which consists in considering the population process killed at an arbitrary time a, which amounts to reflect the contour process below a. The behavior of the contour process below a (and of its local time accumulated below level a) is described by the solution of the corresponding equation reflected below any $a' > a$. The second difficulty comes from the fact that f' is not assumed to be bounded from below. We will prove our convergence result by a combination of Theorem 7 and Girsanov's theorem. For that sake, we shall first consider the case where $|f'|$ is bounded, and then the general case.

7.1 Convergence of $Z^{N,x}$

The aim of this section is to prove the convergence in law as $N \to \infty$ of the two-parameter process $\{Z_t^{N,x},\ t \geq 0, x \geq 0\}$ defined in section 6.5 towards the process $\{Z_t^x,\ t \geq 0, x \geq 0\}$ solution of the SDE (4.10). We need to make precise the topology for which this convergence will hold. We note that the process $Z_t^{N,x}$ (resp., Z_t^x) is a Markov process indexed by x, with values in the space of càdlàg (resp., continuous) functions of t $D([0,\infty); \mathbb{R}_+)$ (resp., $C([0,\infty); \mathbb{R}_+)$). So it will be natural to consider a topology of functions of x, with values in a space of functions of t.

© Springer International Publishing Switzerland 2016

É. Pardoux, *Probabilistic Models of Population Evolution*, Mathematical Biosciences Institute Lecture Series 1.6, DOI 10.1007/978-3-319-30328-4_7

For each fixed x, the process $t \to Z_t^{N,x}$ is càdlàg, constant between its jumps, with jumps of size $\pm N^{-1}$, while the limit process $t \to Z_t^x$ is continuous. On the other hand, both $Z_t^{N,x}$ and Z_t^x are discontinuous as functions of x. The mapping $x \to Z_{\cdot}^x$ has countably many jumps on any compact interval, but the mapping $x \to \{Z_t^x, \ t \geq \varepsilon\}$, where $\varepsilon > 0$ is arbitrary, has finitely many jumps on any compact interval, and it is constant between its jumps. This fact is well-known in the case where f is linear, see section 4.4, and has been proved in the general case in Corollary 2. Recall that $D([0,\infty);\mathbb{R}_+)$, equipped with the distance d_∞^0 defined by (16.4) in [10], is separable and complete, see Theorem 16.3 in [10]. We have the following statement

Theorem 13. *Suppose that Assumption (H1) is satisfied. Then as $N \to \infty$,*

$$\{Z_t^{N,x}, \ t \geq 0, x \geq 0\} \Rightarrow \{Z_t^x, \ t \geq 0, x \geq 0\}$$

in $D([0,\infty);D([0,\infty);\mathbb{R}_+))$, equipped with the Skorokhod topology of the space of càdlàg functions of x, with values in the Polish space $D([0,\infty);\mathbb{R}_+)$ equipped with the metric d_∞^0, where $\{Z_t^x, \ t \geq 0, x \geq 0\}$ is the unique solution of the SDE (4.10).

7.1.1 Tightness of $Z^{N,x}$

Recall (6.3) and (6.5). We first establish a few Lemmas.

Lemma 20. *For all $T > 0$, $x \geq 0$, there exists a constant $C_0 > 0$ such that for all $N \geq 1$,*

$$\sup_{0 \leq t \leq T} \mathbb{E}\left(Z_t^{N,x}\right) \leq C_0.$$

Moreover, for all $t \geq 0$, $N \geq 1$,

$$\mathbb{E}\left(-\int_0^t f(Z_r^{N,x})dr\right) \leq x.$$

PROOF: Let $(\tau_n, n \geq 0)$ be a sequence of stopping times such that τ_n tends to infinity as n goes to infinity and for any n, $\left(M_{t \wedge \tau_n}^{N,x}, t \geq 0\right)$ is a martingale and $Z_{t \wedge \tau_n}^{N,x} \leq n$. Taking the expectation on both sides of equation (6.3) at time $t \wedge \tau_n$, we obtain

$$\mathbb{E}\left(Z_{t \wedge \tau_n}^{N,x}\right) = \frac{\lfloor Nx \rfloor}{N} + \mathbb{E}\left(\int_0^{t \wedge \tau_n} f(Z_r^{N,x})dr\right). \tag{7.1}$$

It follows from the Assumption (H1) on f that

$$\mathbb{E}\left(Z_{t \wedge \tau_n}^{N,x}\right) \leq \frac{\lfloor Nx \rfloor}{N} + \beta \int_0^t \mathbb{E}(Z_{r \wedge \tau_n}^{N,x})dr$$

From Gronwall's and Fatou's Lemmas, we deduce that there exists a constant $C_0 > 0$ which depends only upon x and T such that

$$\sup_{N \geq 1} \sup_{0 \leq t \leq T} \mathbb{E}\left(Z_t^{N,x}\right) \leq C_0.$$

From (7.1), we deduce that

$$-\mathbb{E}\left(\int_0^{t \wedge \tau_n} f(Z_r^{N,x}) dr\right) \leq \frac{\lfloor Nx \rfloor}{N}.$$

Since $-f(Z_r^{N,x}) \geq -\beta Z_r^{N,x}$, the second statement follows using Fatou's Lemma and the first statement. □

We now have the following Lemma.

Lemma 21. *For all $T > 0$, $x \geq 0$, there exists a constant $C_1 > 0$ such that*

$$\sup_{N \geq 1} \mathbb{E}\left(\langle M^{N,x}\rangle_T\right) \leq C_1.$$

PROOF: For any $N \geq 1$ and $k, k' \in \mathbb{Z}_+$ such that $k \leq k'$, we set $z = \frac{k}{N}$ and $z' = \frac{k'}{N}$. We deduce from (6.6) that

$$||f||_{N,z,z'} = \sum_{i=k+1}^{k'} \left\{ 2\left(f(\tfrac{i}{N}) - f(\tfrac{i-1}{N})\right)^+ - \left(f(\tfrac{i}{N}) - f(\tfrac{i-1}{N})\right) \right\}.$$

Hence it follows from Assumption (H1) that

$$||f||_{N,z,z'} \leq 2\beta(z'-z) + f(z) - f(z'). \tag{7.2}$$

We deduce from (7.2), (6.5), and Lemma 20 that

$$\mathbb{E}\left(\langle M^{N,x}\rangle_T\right) \leq \int_0^T \left\{ \left(\sigma^2 + \frac{2\beta}{N}\right) \mathbb{E}(Z_r^{N,x}) - \frac{1}{N}\mathbb{E}\left(f(Z_r^{N,x})\right) \right\} dr$$
$$\leq \left(\sigma^2 + \frac{2\beta}{N}\right) C_0 T + \frac{x}{N}.$$

Hence the Lemma. □

It follows from this that $M^{N,x}$ is in fact a square integrable martingale. We also have

Lemma 22. *For all $T > 0$, $x \geq 0$, there exist two constants $C_2, C_3 > 0$ such that*

$$\sup_{N \geq 1} \sup_{0 \leq t \leq T} \mathbb{E}\left[\left(Z_t^{N,x}\right)^2\right] \leq C_2,$$

$$\sup_{N \geq 1} \sup_{0 \leq t \leq T} \mathbb{E}\left(-\int_0^t Z_r^{N,x} f(Z_r^{N,x}) dr\right) \leq C_3.$$

PROOF: We deduce from (6.3), (A.4), and the fact that $\langle M^{N,x}\rangle_t - [M^{N,x}]_t$ is a local martingale

$$\left(Z_t^{N,x}\right)^2 = \left(\frac{\lfloor Nx \rfloor}{N}\right)^2 + 2\int_0^t Z_r^{N,x} f(Z_r^{N,x})dr + \langle M^{N,x}\rangle_t + M_t^{N,x,(2)}, \qquad (7.3)$$

where $M^{N,x,(2)}$ is a local martingale. Let $(\tau_n, n \geq 1)$ be a sequence of stopping times such that $\lim_{n\to\infty} \tau_n = +\infty$ a.s. and for each $n \geq 1$, $\left(M_{t\wedge\tau_n}^{N,x,(2)}, t \geq 0\right)$ is a martingale. Taking the expectation on the both sides of (7.3) at time $t \wedge \tau_n$ and using Assumption (H1), Lemma 21, and the Gronwall and Fatou Lemmas, we obtain that for all $T > 0$, there exists a constant $C_2 > 0$ such that

$$\sup_{N\geq 1} \sup_{0\leq t\leq T} \mathbb{E}\left(Z_t^{N,x}\right)^2 dr \leq C_2.$$

We also have that

$$2\mathbb{E}\left(-\int_0^{t\wedge\tau_n} Z_r^{N,x} f(Z_r^{N,x})dr\right) \leq \left(\frac{\lfloor Nx \rfloor}{N}\right)^2 + C_1$$

From Assumption (H1), we have $-Z_r^{N,x} f(Z_r^{N,x}) \geq -\beta(Z_r^{N,x})^2$. The second result now follows from Fatou's Lemma. $\qquad\square$

We want to check tightness of the sequence $\{Z^{N,x}, N \geq 0\}$. Because of the very weak assumptions upon f, we cannot use Proposition 37 below. Instead, we now show directly how we can use Aldous' criterion (A), see section A.7. Let $\{\tau_N, N \geq 1\}$ be a sequence of stopping times in $[0,T]$. We deduce from Lemma 22

Proposition 21. *For any $T > 0$ and η, $\varepsilon > 0$, there exists $\delta > 0$ such that*

$$\sup_{N\geq 1} \sup_{0\leq\theta\leq\delta} \mathbb{P}\left(\left|\int_{\tau_N}^{(\tau_N+\theta)\wedge T} f(Z_r^{N,x})dr\right| \geq \eta\right) \leq \varepsilon.$$

PROOF: Let c be a nonnegative constant. Provided $0 \leq \theta \leq \delta$, we have

$$\left|\int_{\tau_N}^{(\tau_N+\theta)\wedge T} f(Z_r^{N,x})dr\right| \leq \sup_{0\leq r\leq c}|f(r)|\delta + \int_{\tau_N}^{(\tau_N+\theta)\wedge T} \mathbf{1}_{\{Z_r^{N,x}>c\}}|f(Z_r^{N,x})|dr$$

But

$$\int_{\tau_N}^{(\tau_N+\theta)\wedge T} \mathbf{1}_{\{Z_r^{N,x}>c\}}|f(Z_r^{N,x})|dr \leq c^{-1}\int_0^T Z_r^{N,x}\left(f^+(Z_r^{N,x}) + f^-(Z_r^{N,x})\right)dr$$

$$\leq c^{-1}\int_0^T \left(2Z_r^{N,x}f^+(Z_r^{N,x}) - Z_r^{N,x}f(Z_r^{N,x})\right)dr$$

$$\leq c^{-1}\int_0^T \left(2\beta(Z_r^{N,x})^2 - Z_r^{N,x}f(Z_r^{N,x})\right)dr.$$

From this and Lemma 22, we deduce that $\forall\, N \geq 1$, again with $\theta \leq \delta$,

$$\sup_{0 \leq \theta \leq \delta} \mathbb{P}\left(\left| \int_{\tau_N}^{(\tau_N + \theta) \wedge T} f(Z_r^{N,x}) dr \right| \geq \eta \right) \leq \eta^{-1} \mathbb{E}\left(\left| \int_{\tau_N}^{(\tau_N + \theta) \wedge T} f(Z_r^{N,x}) dr \right| \right)$$

$$\leq \sup_{0 \leq r \leq c} \frac{|f(r)|\delta}{\eta} + \frac{A}{c\eta},$$

with $A = 2\beta C_2 T + C_3$. The result follows by choosing $c = 2A/\varepsilon\eta$, and then $\delta = \varepsilon\eta/2 \sup_{0 \leq r \leq c} |f(z)|$. $\square$

From Proposition 21, the Lebesgue integral term in the right-hand side of (6.3) satisfies Aldous's condition (A). The same Proposition, Lemma 20, (6.5), and (7.2) imply that $< M^{N,x} >$ satisfies the same condition, hence so does $M^{N,x}$, according to Rebolledo's theorem, see [21]. We have proved

Proposition 22. *For any fixed $x \geq 0$, the sequence of processes $\{Z^{N,x}, N \geq 1\}$ is tight in $D([0,\infty);\mathbb{R}_+)$.*

We deduce from Proposition 22 the following Corollary.

Corollary 6. *For any $0 \leq x < y$ the sequence of processes $\{V^{N,x,y}, N \geq 1\}$ is tight in $D([0,\infty);\mathbb{R}_+)$*

PROOF: For any x fixed the process $Z_t^{N,x}$ has jumps equal to $\pm\frac{1}{N}$ which tend to zero as $N \to \infty$. It follows from this, Proposition 22, and Proposition 37 that any weak limit of a converging subsequence of $Z^{N,x}$ is continuous. We deduce that for any $x, y \geq 0$, the sequence $\{Z^{N,y} - Z^{N,x}, N \geq 1\}$ is tight since $\{Z^{N,x}, N \geq 1\}$ and $\{Z^{N,y}, N \geq 1\}$ are tight and both have a continuous limit as $N \to \infty$. $\square$

7.1.2 Proof of Theorem 13

The next two Propositions will be the main steps in the proof of Theorem 13.

Proposition 23. *For any $n \in \mathbb{N}$, $0 \leq x_1 < x_2 < \cdots < x_n$,*

$$\left(Z^{N,x_1}, Z^{N,x_2}, \cdots, Z^{N,x_n}\right) \Rightarrow \left(Z^{x_1}, Z^{x_2}, \cdots, Z^{x_n}\right)$$

as $N \to \infty$, for the topology of locally uniform convergence in t.

PROOF: We prove the statement in the case $n = 2$ only. The general statement can be proved in a very similar way. For $0 \leq x_1 < x_2$, we consider the process $\left(Z^{N,x_1}, V^{N,x_1,x_2}\right)$, using the notations from section 6.5. The argument preceding the statement of Proposition 22 implies that the sequences of martingales M^{N,x_1} and M^{N,x_1,x_2} are tight. Hence $\left(Z^{N,x_1}, V^{N,x_1,x_2}, M^{N,x_1}, M^{N,x_1,x_2}\right)$ is tight. Thanks to (6.3), (6.5), (6.8), (6.9), and (6.10), any converging subsequence of

$\{Z^{N,x_1}, V^{N,x_1,x_2}, M^{N,x_1}, M^{N,x_1,x_2}, N \geq 1\}$ has a weak limit $(Z^{x_1}, V^{x_1,x_2}, M^{x_1}, M^{x_1,x_2})$ which satisfies

$$Z_t^{x_1} = x_1 + \int_0^t f(Z_s^{x_1})ds + M_t^{x_1}$$

$$V_t^{x_1,x_2} = x_2 - x_1 + \int_0^t [f(Z_s^{x_1} + V_s^{x_1,x_2}) - f(Z_s^{x_1})]ds + M_t^{x_1,x_2},$$

where the continuous martingales M^{x_1} and M^{x_1,x_2} satisfy

$$\langle M^x \rangle_t = \sigma^2 \int_0^t Z_s^{x_1} ds, \ \langle M^{x_1,x_2} \rangle_t = \sigma^2 \int_0^t V_s^{x_1,x_2} ds, \ \langle M^{x_1}, M^{x_1,x_2} \rangle_t = 0.$$

This implies that the pair (Z^{x_1}, V^{x_1,x_2}) is a weak solution of the system of SDEs (4.10) and (4.18), driven by the same space-time white noise. The result follows from the uniqueness of the system, see Theorem 5. □

Proposition 24. *There exists a constant C, which depends only upon θ and T, such that for any $0 \leq x < y < z$, which are such that $y - x \leq 1$, $z - y \leq 1$,*

$$\mathbb{E}\left[\sup_{0 \leq t \leq T} |Z_t^{N,y} - Z_t^{N,x}|^2 \times \sup_{0 \leq t \leq T} |Z_t^{N,z} - Z_t^{N,y}|^2\right] \leq C|z - x|^2.$$

We first prove the

Lemma 23. *For any $0 \leq x < y$, we have*

$$\sup_{0 \leq t \leq T} \mathbb{E}\left(Z_t^{N,y} - Z_t^{N,x}\right) = \sup_{0 \leq t \leq T} \mathbb{E}(V_t^{N,x,y}) \leq \left(\frac{\lfloor Ny \rfloor}{N} - \frac{\lfloor Nx \rfloor}{N}\right)e^{\beta T},$$

PROOF: Let $(\tau_n, n \geq 0)$ be a sequence of stopping times such that $\lim_{n \to \infty} \tau_n = +\infty$ and for each $n \geq 1$, $\left(M_{t \wedge \tau_n}^{N,x,y}, t \geq 0\right)$ is a martingale. Taking the expectation on both sides of (6.8) at time $t \wedge \tau_n$, we obtain that

$$\mathbb{E}(V_{t \wedge \tau_n}^{N,x,y}) \leq \left(\frac{\lfloor Ny \rfloor}{N} - \frac{\lfloor Nx \rfloor}{N}\right) + \beta \int_0^t \mathbb{E}(V_{r \wedge \tau_n}^{N,x,y})dr \qquad (7.4)$$

Using Gronwall's and Fatou's Lemmas, we obtain that

$$\sup_{0 \leq t \leq T} \mathbb{E}(V_t^{N,x,y}) \leq \left(\frac{\lfloor Ny \rfloor}{N} - \frac{\lfloor Nx \rfloor}{N}\right)e^{\beta T}.$$

□

PROOF OF PROPOSITION 24 From equation (6.8), using a stopping time argument as above, Lemma 23, and Fatou's Lemma, where we take advantage of the inequality $f(Z_r^{N,x}) - f(Z_r^{N,x} + V_r^{N,x,y}) \geq -\beta V_r^{N,x,y}$, we deduce that

$$\mathbb{E}\left(\int_0^t \left[f(Z_r^{N,x}) - f(Z_r^{N,x} + V_r^{N,x,y})\right] dr\right) \le \frac{\lfloor Ny \rfloor}{N} - \frac{\lfloor Nx \rfloor}{N}. \tag{7.5}$$

We now deduce from (6.9), Lemma 23, and inequalities (7.5) and (7.2) that for each $t > 0$, there exists a constant $C(t) > 0$ such that

$$\mathbb{E}\left(\langle M^{N,x,y}\rangle_t\right) \le C(t)\left(\frac{\lfloor Ny \rfloor}{N} - \frac{\lfloor Nx \rfloor}{N}\right). \tag{7.6}$$

This implies that $M^{N,x,y}$ is in fact a square integrable martingale. For any $0 \le x < y < z$, we have $Z_t^{N,z} - Z_t^{N,y} = V_t^{N,y,z}$ and $Z_t^{N,y} - Z_t^{N,x} = V_t^{N,x,y}$ for any $t \ge 0$. On the other hand we deduce from (6.8) and Assumption (H1) that

$$\sup_{0 \le t \le T} (V_t^{N,x,y})^2 \le 3\left(\frac{\lfloor Ny \rfloor}{N} - \frac{\lfloor Nx \rfloor}{N}\right)^2 + 3\beta^2 T \int_0^T \sup_{0 \le s \le r} (V_s^{N,x,y})^2 dr$$
$$+ 3 \sup_{0 \le t \le T} \left(M_t^{N,x,y}\right)^2$$

and

$$\sup_{0 \le t \le T} (V_t^{N,y,z})^2 \le 3\left(\frac{\lfloor Nz \rfloor}{N} - \frac{\lfloor Ny \rfloor}{N}\right)^2 + 3\beta^2 T \int_0^t \sup_{0 \le s \le r} (V_s^{N,y,z})^2 dr$$
$$+ 3 \sup_{0 \le t \le T} \left(M_t^{N,y,z}\right)^2.$$

Now let $\mathcal{G}^{x,y} := \sigma\left(Z_t^{N,x}, Z_t^{N,y}, t \ge 0\right)$ be the filtration generated by $Z^{N,x}$ and $Z^{N,y}$. It is clear that for any t, $V_t^{N,x,y}$ is measurable with respect to $\mathcal{G}^{x,y}$. We then have

$$\mathbb{E}\left[\sup_{0 \le t \le T} |V_t^{N,x,y}|^2 \times \sup_{0 \le t \le T} |V_t^{N,y,z}|^2\right] = \mathbb{E}\left[\sup_{0 \le t \le T} |V_t^{N,x,y}|^2 \mathbb{E}\left(\sup_{0 \le t \le T} |V_t^{N,y,z}|^2 \big| \mathcal{G}^{x,y}\right)\right].$$

Conditionally upon $Z^{N,x}$ and $Z^{N,y} = u(.)$, $V^{N,y,z}$ solves the following SDE

$$V_t^{N,y,z} = \frac{\lfloor Nz \rfloor - \lfloor Ny \rfloor}{N} + \int_0^t \left[f(V_r^{N,y,z} + u(r)) - f(u(r))\right] dr + M_t^{N,y,z},$$

where $M^{N,y,z}$ is a martingale conditionally upon $\mathcal{G}^{x,y}$, hence the arguments used in Lemma 23 lead to

$$\sup_{0 \le t \le T} \mathbb{E}\left(V_t^{N,y,z} | \mathcal{G}^{x,y}\right) \le \left(\frac{\lfloor Nz \rfloor}{N} - \frac{\lfloor Ny \rfloor}{N}\right) e^{\beta T},$$

and those used to prove (7.5) yield

$$\mathbb{E}\left(\int_0^t f(Z_r^{N,y}) - f(Z_r^{N,y} + V_r^{N,y,z}) dr | \mathcal{G}^{x,y}\right) \le \frac{\lfloor Nz \rfloor}{N} - \frac{\lfloor Ny \rfloor}{N}.$$

From this we deduce (see the proof of (7.6)) that

$$\mathbb{E}\left(\langle M^{N,y,z}\rangle_t|\mathscr{G}^{x,y}\right) \le C(t)\left(\frac{\lfloor Nz\rfloor}{N} - \frac{\lfloor Ny\rfloor}{N}\right).$$

From Doob's inequality we have

$$\mathbb{E}\left(\sup_{0\le t\le T}|M_t^{N,y,z}|^2|\mathscr{G}^{x,y}\right) \le 4\mathbb{E}\left(\langle M^{N,y,z}\rangle_T|\mathscr{G}^{x,y}\right)$$

$$\le C(T)\left(\frac{\lfloor Nz\rfloor}{N} - \frac{\lfloor Ny\rfloor}{N}\right).$$

Since $0 < z - y < 1$, we deduce that

$$\mathbb{E}\left(\sup_{0\le t\le T}|V_t^{N,y,z}|^2|\mathscr{G}^{x,y}\right) \le 3(1+C(T))\left(\frac{\lfloor Nz\rfloor}{N} - \frac{\lfloor Ny\rfloor}{N}\right)$$
$$+ 3\beta^2 T\int_0^T \mathbb{E}\left(\sup_{0\le s\le r}(V_s^{N,y,z})^2|\mathscr{G}^{x,y}\right)dr,$$

From this and Gronwall's Lemma we deduce that there exists a constant $K_1 > 0$ such that

$$\mathbb{E}\left(\sup_{0\le t\le T}|V_t^{N,y,z}|^2|\mathscr{G}^{x,y}\right) \le K_1\left(\frac{\lfloor Nz\rfloor}{N} - \frac{\lfloor Ny\rfloor}{N}\right). \tag{7.7}$$

Similarly we have

$$\mathbb{E}\left[\sup_{0\le t\le T}(V_s^{N,x,y})^2\right] \le K_1\left(\frac{\lfloor Ny\rfloor}{N} - \frac{\lfloor Nx\rfloor}{N}\right).$$

Since $0 \le y - x < z - x$ and $0 \le z - y < z - x$, we deduce that

$$\mathbb{E}\left[\sup_{0\le t\le T}|V_t^{N,x,y}|^2 \times \sup_{0\le t\le T}|V_t^{N,y,z}|^2\right] \le K_1^2\left(\frac{\lfloor Nz\rfloor}{N} - \frac{\lfloor Nx\rfloor}{N}\right)^2,$$

hence the result.

PROOF OF THEOREM 13 We now show that for any $T > 0$,

$$\{Z_t^{N,x}, 0\le t\le T, x\ge 0\} \Rightarrow \{Z_t^x, 0\le t\le T, x\ge 0\}$$

in $D([0,\infty); D([0,T],\mathbb{R}_+))$. From Theorems 13.1 and 16.8 in [10], since from Proposition 23, for all $n\ge 1, 0 < x_1 < \cdots < x_n$,

$$(Z_{\cdot}^{N,x_1},\ldots,Z_{\cdot}^{N,x_n}) \Rightarrow (Z_{\cdot}^{x_1},\ldots,Z_{\cdot}^{x_n})$$

in $D([0,T];\mathbb{R}^n)$, it suffices to show that for all $\bar{x} > 0$, ε, $\eta > 0$, there exists $N_0 \ge 1$ and $\delta > 0$ such that for all $N \ge N_0$,

$$\mathbb{P}(w_{\bar{x},\delta}(Z^N) \geq \varepsilon) \leq \eta, \qquad (7.8)$$

where for a function $(x,t) \to z(x,t)$

$$w_{\bar{x},\delta}(z) = \sup_{0 \leq x_1 \leq x \leq x_2 \leq \bar{x}, x_2 - x_1 \leq \delta} \inf\{\|z(x,\cdot) - z(x_1,\cdot)\|, \|z(x_2,\cdot) - z(x,\cdot)\|\},$$

with the notation $\|z(x,\cdot)\| = \sup_{0 \leq t \leq T} |z(x,t)|$. But from the proof of Theorem 13.5 in [10], (7.8) for Z^N follows from Proposition 24. $\qquad\square$

7.2 Convergence of the Contour Process H^N

In this section, we assume that $f \in C^1$. In this case, Assumption (H1) is equivalent to

Assumption (H1') There exists a constant $\beta > 0$ such that for all $x \geq 0$, $f'(x) \leq \beta$,
  which we assume to be in force in this section.

7.2.1 The Case Where f' Is Bounded

We assume in this subsection that $|f'(x)| \leq C$ for all $x \geq 0$ and some $C > 0$. This constitutes the first step of the proof of convergence of H^N.
  As explained at the end of section 6.6, in this case we can use Girsanov's theorem to bring us back to the situation studied in section 5.3.
  Recalling equations (6.13) and (6.15), we note that

$$H_s^{a,N} = \mathscr{M}_s^{1,a,N} - \mathscr{M}_s^{2,a,N} + 2^{-1}[L_s^{a,N}(0) - L_s^{a,N}(a^-)] + \varepsilon_N, \text{ where}$$
$$\varepsilon_N = (4N)^{-1}(1 - V_s^{a,N}) - 2^{-1}L_{0^+}^{a,N}(0).$$

Moreover, from (6.12), (6.14), and (6.15),

$$[\mathscr{M}^{1,a,N}]_s = \frac{4}{N^2\sigma^4}Q_s^{1,a,N}, \quad [\mathscr{M}^{2,a,N}]_s = \frac{4}{N^2\sigma^4}Q_s^{2,a,N}$$

$$\langle \mathscr{M}^{1,a,N}\rangle_s = \frac{4}{\sigma^2}\int_0^s \mathbf{1}_{V_r^N=-1}dr, \ \langle \mathscr{M}^{2,a,N}\rangle_s = \frac{4}{\sigma^2}\int_0^s \mathbf{1}_{V_r^N=1}dr,$$

$$[Y^{a,N}]_s = \frac{4}{N^2\sigma^4}\int_0^s |Y_{r^-}^{a,N}|^2 \left[\left|(f_N')^+ \left(\frac{\sigma^2}{4}L_{r^-}^{a,N}(H_r^{a,N})\right)\right|^2 dQ_r^{1,a,N}\right.$$
$$\left. + \left|(f_N')^- \left(\frac{\sigma^2}{4}L_{r^-}^{a,N}(H_r^{a,N})\right)\right|^2 dQ_r^{2,a,N}\right]$$

$$\langle Y^{a,N}\rangle_s = \frac{4}{\sigma^2}\int_0^s |Y_r^{a,N}|^2 \Big[\Big|(f_N')^+\Big(\frac{\sigma^2}{4}L_r^{a,N}(H_r^{a,N})\Big)\Big|^2 \mathbf{1}_{V_r^{a,N}=-1}$$

$$+ \Big|(f_N')^-\Big(\frac{\sigma^2}{4}L_r^{a,N}(H_r^{a,N})\Big)\Big|^2 \mathbf{1}_{V_r^{a,N}=1}\Big]dr$$

$$[Y^{a,N},\mathcal{M}^{1,a,N}]_s = \frac{4}{N^2\sigma^4}\int_0^s Y_{r^-}^{a,N}(f_N')^+\Big(\frac{\sigma^2}{4}L_{r^-}^{a,N}(H_r^{a,N})\Big)dQ_r^{1,a,N}$$

$$[Y^{a,N},\mathcal{M}^{2,a,N}]_s = \frac{4}{N^2\sigma^4}\int_0^s Y_{r^-}^{a,N}(f_N')^-\Big(\frac{\sigma^2}{4}L_{r^-}^{a,N}(H_r^{a,N})\Big)dQ_r^{2,a,N}$$

$$\langle Y^{a,N},\mathcal{M}^{1,a,N}\rangle_s = \frac{4}{\sigma^2}\int_0^s Y_r^{a,N}(f_N')^+\Big(\frac{\sigma^2}{4}L_r^{a,N}(H_r^{a,N})\Big)\mathbf{1}_{V_r^{a,N}=-1}dr$$

$$\langle Y^{a,N},\mathcal{M}^{2,a,N}\rangle_s = \frac{4}{\sigma^2}\int_0^s Y_r^{a,N}(f_N')^-\Big(\frac{\sigma^2}{4}L_r^{a,N}(H_r^{a,N})\Big)\mathbf{1}_{V_r^{a,N}=1}dr,$$

while

$$[\mathcal{M}^{1,a,N},\mathcal{M}^{2,a,N}]_s = \langle\mathcal{M}^{1,a,N},\mathcal{M}^{1,a,N}\rangle_s = 0.$$

Recall Corollary 3 and Lemma 15. Since f' is bounded, the same is true for $(f')_N(x) = N[f(x+1/N)-f(x)]$, uniformly with respect to N. It is not difficult to deduce from the above formulae and Proposition 37 that $\{(H^{a,N},\mathcal{M}^{1,a,N},\mathcal{M}^{2,a,N},Y^{a,N}),\ N\geq 1\}$ is a tight sequence in $C([0,\infty))\times D([0,+\infty))^3$. Hence at least along a subsequence (but we do not distinguish between the notation for the subsequence and for the sequence),

$$(H^{a,N},\mathcal{M}^{1,a,N},\mathcal{M}^{2,a,N},Y^{a,N})\Rightarrow(H^a,\mathcal{M}^1,\mathcal{M}^2,Y^a)$$

as $N\to\infty$ in $C([0,\infty))\times D([0,+\infty))^3$, the limit being continuous (since the jumps of $\mathcal{M}^{1,a,N}$, $\mathcal{M}^{2,a,N}$, and $Y^{a,N}$ tend to zero). Moreover

$$\langle Y^{a,N}\rangle_s \Rightarrow \frac{2}{\sigma^2}\int_0^s |Y_r^a|^2\times\Big|f'\Big(\frac{\sigma^2}{4}L_r(H_r)\Big)\Big|^2 dr$$

$$\langle\mathcal{M}^{1,a,N}\rangle_s \Rightarrow \frac{2}{\sigma^2}s,$$

$$\langle\mathcal{M}^{2,a,N}\rangle_s \Rightarrow \frac{2}{\sigma^2}s,$$

$$\langle Y^{a,N},\mathcal{M}^{1,a,N}\rangle_s \Rightarrow \frac{2}{\sigma^2}\int_0^s Y_r^a f'^+\Big(\frac{\sigma^2}{4}L_r(H_r)\Big)dr,$$

$$\langle Y^{a,N},\mathcal{M}^{2,a,N}\rangle_s \Rightarrow \frac{2}{\sigma^2}\int_0^s Y_r^a f'^-\Big(\frac{\sigma^2}{4}L_r(H_r)\Big)dr.$$

It follows from the above that Corollary 3 can be enriched as follows

Proposition 25. *For each $a > 0$, as $N \to \infty$,*

$$\left(H^{a,N}, M^{1,a,N}, M^{2,a,N}, L^{a,N}_\cdot(0), L^{a,N}_\cdot(a^-), Y^{a,N}\right) \Longrightarrow \left(H^a, \frac{\sqrt{2}}{\sigma}B^1, \frac{\sqrt{2}}{\sigma}B^2, L^a_\cdot(0), L^a_\cdot(a^-), Y^a\right),$$

where B^1 and B^2 are two mutually independent standard Brownian motions, $L^a_\cdot(0)$ (resp., $L^a_\cdot(a^-)$) denotes the local time of the continuous semimartingale H^a at level 0 (resp., at level a^-). Moreover

$$H^a_s = \frac{\sqrt{2}}{\sigma}(B^1_s - B^2_s) + \frac{1}{2}[L^a_s(0) - L^a_s(a^-)], \quad and$$

$$Y^a_s = 1 + \frac{\sqrt{2}}{\sigma}\int_0^s Y^a_r \left[f'^+\left(\frac{\sigma^2}{4}L_r(H_r)\right)dB^1_r + f'^-\left(\frac{\sigma^2}{4}L_r(H_r)\right)dB^2_r\right].$$

We clearly have

$$Y^a_s = \exp\left(\frac{\sqrt{2}}{\sigma}\int_0^s \left[f'^+\left(\frac{\sigma^2}{4}L_r(H_r)\right)dB^1_r + f'^-\left(\frac{\sigma^2}{4}L_r(H_r)\right)dB^2_r\right]\right.$$
$$\left. - \frac{1}{\sigma^2}\int_0^s \left|f'\left(\frac{\sigma^2}{4}L_r(H_r)\right)\right|^2 dr\right). \tag{7.9}$$

Since f' is bounded, it is plain that $\mathbb{E}(Y^a_s) = 1$ for all $s > 0$. Let now $\tilde{\mathbb{P}}^a$ denote the probability measure such that

$$\left.\frac{d\tilde{\mathbb{P}}^a}{d\mathbb{P}}\right|_{\mathscr{G}^a_s} = Y^a_s, \tag{7.10}$$

where $\mathscr{G}^a_s := \sigma\{H^a_r,\ 0 \le r \le s\}$. It follows from Girsanov's theorem (see Proposition 35 below) that there exist two mutually independent standard $\tilde{\mathbb{P}}^a$-Brownian motions $\tilde{B}^1$ and $\tilde{B}^2$ such that

$$B^1_s = \frac{\sqrt{2}}{\sigma}\int_0^s f'^+\left(\frac{\sigma^2}{4}L_r(H_r)\right)dr + \tilde{B}^1_s,$$

$$B^2_s = \frac{\sqrt{2}}{\sigma}\int_0^s f'^-\left(\frac{\sigma^2}{4}L_r(H_r)\right)dr + \tilde{B}^2_s.$$

Consequently

$$\frac{\sqrt{2}}{\sigma}(B^1_s - B^2_s) = \frac{2}{\sigma}B_s + \frac{2}{\sigma^2}\int_0^s f'\left(\frac{\sigma^2}{4}L_r(H_r)\right)dr,$$

where $B_s = (\sqrt{2})^{-1}(\tilde{B}^1_s - \tilde{B}^2_s)$ is a standard Brownian motion under $\tilde{\mathbb{P}}^a$. Consequently H^a is a weak solution of the SDE

$$H_s^a = \frac{2}{\sigma^2} \int_0^s f'\left(\frac{\sigma^2}{4} L_r(H_r)\right) dr + \frac{2}{\sigma} B_s + \frac{1}{2}[L_s^a(0) - L_s^a(a^-)], \qquad (7.11)$$

where $L_s^a(t)$ denotes the local time accumulated at level t up to time s by the process H^a.

We note that under $\widetilde{\mathbb{P}}^{a,N}$, $Y^{a,N}$ solves (6.13), and under $\widetilde{\mathbb{P}}^a$, Y^a solves the SDE (7.11).

Let us establish a general Lemma

Lemma 24. *Let* (ξ_N, η_N), (ξ, η) *be random pairs defined on a probability space* $(\Omega, \mathscr{F}, \mathbb{P})$, *with* η_N, η *nonnegative scalar random variables, and* ξ_N, ξ *taking values in some complete separable metric space* $\mathscr{X}$. *Assume that* $\mathbb{E}[\eta_N] = \mathbb{E}[\eta] = 1$. *Write* $(\tilde{\xi}_N, \tilde{\eta}_N)$ *for the random pair* (ξ_N, η_N) *defined under the probability measure* $\widetilde{\mathbb{P}}^N$ *which has density* η_N *with respect to* $\mathbb{P}$, *and* $(\tilde{\xi}, \tilde{\eta})$ *for the random pair* (ξ, η) *defined under the probability measure* $\widetilde{\mathbb{P}}$ *which has the density* η *with respect to* $\mathbb{P}$. *If* (ξ_N, η_N) *converges in distribution to* (ξ, η), *then* $(\tilde{\xi}_N, \tilde{\eta}_N)$ *converges in distribution to* $(\tilde{\xi}, \tilde{\eta})$.

PROOF: Due to the equality $\mathbb{E}[\eta_N] = \mathbb{E}[\eta] = 1$ and Scheffé's theorem (see Theorem 16.12 in [9]), the sequence η_N is uniformly integrable. Hence for all bounded continuous $F : \mathscr{X} \times \mathbb{R}_+ \to \mathbb{R}$,

$$\mathbb{E}[F(\tilde{\xi}_N, \tilde{\eta}_N)] = \mathbb{E}[F(\xi_N, \eta_N)\eta_N] \to \mathbb{E}[F(\xi, \eta)\eta] = \mathbb{E}[F(\tilde{\xi}, \tilde{\eta})].$$

$\square$

It follows readily from Proposition 25 and Lemma 24

Proposition 26. *For any* $a > 0$, *as* $N \to \infty$, $H^{a,N}$, *solution of* (6.13) *where the intensity of* P^N *is* $\sigma^2 N^2$, *converges in law towards the solution* H^a *of the SDE* (7.11).

We now define for each $a, x > 0$ the stopping time

$$\tau_x^a = \inf\left\{s > 0, \; L_s^a(0) > \frac{4}{\sigma^2} x\right\}.$$

Combining the above arguments with those of Proposition 18, we deduce that

Lemma 25. *For any* $k \geq 1$, $0 < x_1 < x_2 < \cdots < x_k$, $a > 0$, *as* $N \to \infty$,

$$(H^{a,N}, \tau_{x_1}^{a,N}, \tau_{x_2}^{a,N}, \ldots, \tau_{x_k}^{a,N}, Y^{a,N}) \Rightarrow (H^a, \tau_{x_1}^a, \tau_{x_2}^a, \ldots, \tau_{x_k}^a, Y^a)$$

weakly in $C(\mathbb{R}_+; \mathbb{R}_+) \times \mathbb{R}_+^k \times C(\mathbb{R}_+; \mathbb{R}_+)$.

We can now prove an extension of the Ray–Knight theorem

Proposition 27. *Assume that* f' *is bounded. Then for any* $a > 0$, *the process*

$$\left\{\frac{\sigma^2}{4} L_{\tau_x^a}^a(t), \; 0 \leq t < a, \; x > 0\right\}$$

is a weak solution of equation (4.10) *on the time interval* $[0, a)$.

PROOF: Fix an arbitrary integer $k \geq 1$ and let $0 < x_1 < x_2 < \cdots < x_k$, $g_1, g_2, \ldots, g_k \in C([0,a]; \mathbb{R})$. It follows from Corollary 5 that we have the identity in law

$$\left(\int_0^a g_1(t) Z_t^{N,x_1} dt, \ldots, \int_0^a g_k(t) Z_t^{N,x_k} dt \right)$$

$$\stackrel{(d)}{=} \left(\frac{\sigma^2}{4} \int_0^a g_1(t) L_{\tau_{x_1}^{a,N}}^{a,N}(t) dt, \ldots, \frac{\sigma^2}{4} \int_0^a g_k(t) L_{\tau_{x_k}^{a,N}}^{a,N}(t) dt \right)$$

$$= \left(\int_0^{\tau_{x_1}^{a,N}} g_1(H_r^{a,N}) dr, \ldots, \int_0^{\tau_{x_k}^{a,N}} g_k(H_r^{a,N}) dr \right),$$

where the second equality follows from the "occupation times formula" for zigzag curves, see (5.15). It follows from Proposition 23 that the term on the left converges in law as $N \to \infty$ towards

$$\left(\int_0^a g_1(t) Z_t^{x_1} dt, \ldots, \int_0^a g_k(t) Z_t^{x_k} dt \right),$$

while Lemma 25 implies that the last term on the right converges to

$$\left(\int_0^{\tau_{x_1}^a} g_1(H_r^a) dr, \ldots, \int_0^{\tau_{x_k}^a} g_k(H_r^a) dr \right)$$

$$= \left(\frac{\sigma^2}{4} \int_0^a g_1(t) L_{\tau_{x_1}^a}^a(t) dt, \ldots, \frac{\sigma^2}{4} \int_0^a g_k(t) L_{\tau_{x_k}^a}^a(t) dt \right),$$

where the last identity follows from the occupation times formula (see Proposition 34 below). Consequently for any $k \geq 1$, $0 < x_1 < \cdots < x_k$, $g_1, \ldots, g_k \in C([0,a]; \mathbb{R})$,

$$\left(\int_0^a g_1(t) Z_t^{x_1} dt, \ldots, \int_0^a g_k(t) Z_t^{x_k} dt \right) \stackrel{(d)}{=} \left(\frac{\sigma^2}{4} \int_0^a g_1(t) L_{\tau_{x_1}^a}^a(t) dt, \ldots, \frac{\sigma^2}{4} \int_0^a g_k(t) L_{\tau_{x_k}^a}^a(t) dt \right),$$

which implies the result, since Z is the unique solution of (4.10). □

7.2.2 The General Case ($f \in C^1$ and $f' \leq \beta$)

Y^a is still defined by (7.9). However, it is not clear a priori that $\mathbb{E}[Y_s^a] = 1$ for all $s > 0$ and we need to justify the fact that we can apply Girsanov's theorem.

For each $x > 0$, $a > 0$, and $n \geq 1$, let

$$T_n^a = \inf\{s > 0, \sup_{0 \leq t < a} L_s^a(t) > n\}.$$

It is plain that the process $f'(L_r^a(H_r^a))$ is bounded on the random interval $[0, T_n^a]$, hence $\mathbb{E}[Y_{s \wedge T_n^a}^a] = 1$ for all $s > 0$, and from Proposition 35 below that we can define

$\tilde{\mathbb{P}}^a$ on $\cup_n \mathscr{F}_{T_n^a}$, which is a probability on each $\mathscr{F}_{T_n^a}$, by

$$\left. \frac{d\tilde{\mathbb{P}}^a}{d\mathbb{P}} \right|_{\mathscr{F}_{T_n^a}} = Y_{T_n^a}^a . \tag{7.12}$$

We now establish

Lemma 26. *For any* $x > 0$, $a > 0$, $\mathbb{P}(T_n^a < \tau_x^a) \to 0$, *and* $\tilde{\mathbb{P}}^a(T_n^a < \tau_x^a) \to 0$, *as* $n \to \infty$.

PROOF: It follows from Theorem 8 that

$$\mathbb{P}(T_n^a < \tau_x^a) = \mathbb{P}(\sup_{0 \le t < a} X_t^x > n),$$

where X_t^x is critical Feller diffusion, solution of the SDE

$$X_t^x = x + 2 \int_0^t \sqrt{X_r^x} dB_r.$$

But from Doob's inequality and Gronwall's Lemma,

$$\mathbb{E}\left[\sup_{0 \le r \le t} (X_r^x)^2 \right] \le 2x^2 + 4\mathbb{E}\left(\sup_{0 \le r \le t} \left| \int_0^r \sqrt{X_s^x} dB_s \right|^2 \right)$$

$$\le 2x^2 + 16\mathbb{E}(\int_0^t X_r^x dr)$$

$$\le 2x^2 + 16tx.$$

Hence

$$\mathbb{P}(T_n^a < \tau_x^a) \le \frac{\mathbb{E}\left[\left(\sup_{0 \le t \le a} X_t^x \right)^2 \right]}{n^2}$$

$$\le \frac{2x^2 + 16ax}{n^2},$$

which tends to 0 as $n \to \infty$.

Now let $f_n \in C_b^1(\mathbb{R})$ be such that $f_n(z) = f(z)$, for any $0 \le z \le n$. Applying Proposition 27 with f_n, and noting that on the random interval $[0, T_n^a]$, $f_n'(L_s^a(H_s^a)) = f'(L_s^a(H_s^a))$, we have that

$$\tilde{\mathbb{P}}^a(T_n^a < \tau_x^a) = \mathbb{P}(\sup_{0 \le t < a} Z_t^x > n),$$

where Z_t^x solves the SDE

$$Z_t^x = x + \int_0^t f(Z_r^x)dr + 2\int_0^t \sqrt{Z_r^x}dB_r.$$

Now since $f(z) \le \beta z$, $Z_t^x \le Y_t^x$, solution of the SDE

$$Y_t^x = x + \int_0^t \beta Y_r^x dr + 2\int_0^t \sqrt{Y_r^x}dB_r.$$

A slight extension of the above argument shows that for some constant $C(x,\beta,a)$,

$$\tilde{\mathbb{P}}^a(T_n^a < \tau_x^a) \le \frac{C(x,\beta,a)}{n^2}.$$

□

We can now prove

Proposition 28. $\tilde{\mathbb{P}}^a$ *being defined by (7.12), we have that $\tilde{\mathbb{P}}^a \ll \mathbb{P}$ on $\mathscr{F}_{\tau_x^a}^a$ for any* $x > 0$, *and moreover*

$$\frac{d\tilde{\mathbb{P}}^a}{d\mathbb{P}}|_{\mathscr{F}_{\tau_x^a}^a} = Y_{\tau_x^a}^a.$$

PROOF: For any $A \in \mathscr{F}_{\tau_x^a}^a$, $A \cap \{\tau_x^a \le T_n^a\} \in \mathscr{F}_{T_n^a \wedge \tau_x^a}^a \subset \mathscr{F}_{T_n^a}^a$,

$$\tilde{\mathbb{P}}^a(A \cap \{\tau_x^a \le T_n^a\}) = \int_{A \cap \{\tau_x^a \le T_n^a\}} Y_{T_n^a \wedge \tau_x^a}^a d\mathbb{P}$$

$$= \int_{A \cap \{\tau_x^a \le T_n^a\}} Y_{\tau_x^a}^a d\mathbb{P}.$$

Taking the limit as $n \to \infty$ in this identity with the help of Lemma 26 and the monotone convergence theorem, we deduce that

$$\tilde{\mathbb{P}}^a(A) = \int_A Y_{\tau_x^a}^a d\mathbb{P}.$$

□

We can now extend Proposition 27 to our standard assumptions.

Proposition 29. *Assume that f satisfies Assumption (H1'). Then for any $a > 0$, the process*

$$\left\{ \frac{\sigma^2}{4} L_{\tau_x^a}^a(t), \ 0 \le t < a, \ x > 0 \right\}$$

is a weak solution of equation (4.10) on the time interval $[0,a)$.

PROOF: Consider a sequence $\{f_n,\ n \geq 1\} \subset C_b^1(\mathbb{R}_+)$, as introduced in the proof of Lemma 26. Let $Z^{n,x}$, $H^{n,a}$, and $L^{n,a}$ denote the corresponding population process, contour process, and its local time. From Proposition 27 follows the identity in law

$$\left\{ \frac{\sigma^2}{4} L_{\tau_x^a}^{n,a}(t),\ 0 \leq t < a,\ x > 0 \right\} \overset{(d)}{=} \{Z_t^{n,x},\ 0 \leq t < a,\ x > 0\}.$$

For each $x > 0$, both $\{L_{\tau_x^a}^{n,a}(t),\ 0 \leq t < a,\ 0 < x' \leq x\}$ and $\{Z_t^{n,x},\ 0 \leq t < a,\ 0 < x' \leq x\}$ converge a.s. towards $\{L_{\tau_x^a}^{a}(t),\ 0 \leq t < a,\ 0 < x' \leq x\}$ and $\{Z_t^{x},\ 0 \leq t < a,\ 0 < x' \leq x\}$ (which are associated with the original function f), in the sense that the set where the sequence equals its limit increases a.s. to Ω as $n \to \infty$, as a consequence of Lemma 26. The result follows, since $x > 0$ is arbitrary. $\qquad\square$

Chapter 8
Continuous Model with Interaction

The first goal of this chapter is to describe the genealogy of the continuous population model with interaction, whose size evolves according to the SDE (4.10)

$$Z_t^x = x + \int_0^t f(Z_s^x)\,ds + \sigma \int_0^t \int_0^{Z_s^x} W(ds, du),$$

i.e., to prove a generalization of the second Ray–Knight theorem adapted to this SDE. This is a rather immediate consequence of the results of section 7.2, where we just have to play in the supercritical case with the fact that the level of reflection a is arbitrary. The second part of this chapter is the continuous state counterpart of section 6.4. It answers the question whether or not, as an effect of the interaction, the extinction time and the total mass of the genealogical forest of trees remain finite in the limit of an infinite mass of ancestors at time 0.

8.1 Genealogy in Continuous Population with Interaction

In this section we suppose that f satisfies Assumption (H1'). For each $a > 0$, $\widetilde{\mathbb{P}}^a$ is the law of the unique weak the solution of the reflected SDE

$$H_s^a = \frac{2}{\sigma^2} \int_0^s f'\left(\frac{\sigma^2}{4} L_r^a(H_r^a)\right) dr + \frac{2}{\sigma} B_s + \frac{1}{2}[L_s^a(0) - L_s^a(a^-)].$$

For any $0 < a < b$, both $\widetilde{\mathbb{P}}^{N,a} \Rightarrow \widetilde{\mathbb{P}}^a$ and $\widetilde{\mathbb{P}}^{N,b} \Rightarrow \widetilde{\mathbb{P}}^b$. It then follows from Lemma 19 and the continuity of the mapping $\Pi^{a,b}$ that

$$\widetilde{\mathbb{P}}^a = \widetilde{\mathbb{P}}^b (\Pi^{a,b})^{-1}.$$

© Springer International Publishing Switzerland 2016
É. Pardoux, *Probabilistic Models of Population Evolution*, Mathematical Biosciences Institute Lecture Series 1.6, DOI 10.1007/978-3-319-30328-4_8

Hence

$$\{L^b_{\tau^b_x}(t),\ 0 \le t < a, x > 0\} \overset{(d)}{=} \{L^a_{\tau^a_x}(t),\ 0 \le t < a, x > 0\},$$

and, as in section 5.4, we can define a projective limit, which is a process $\{\mathscr{L}_x(t),\ t \ge 0, x > 0\}$ such that for each $a > 0$,

$$\{\mathscr{L}_x(t),\ 0 \le t < a, x > 0\} \overset{(d)}{=} \{L^a_{\tau^a_x}(t),\ 0 \le t < a, x > 0\}.$$

It follows readily from Proposition 29

Theorem 14. *Assume that f satisfies Assumption (H1'). Then the process*

$$\left\{ \frac{\sigma^2}{4} \mathscr{L}_x(t),\ t \ge 0,\ x > 0 \right\}$$

is a weak solution of equation (4.10).

Consider now the SDE

$$H_s = \frac{2}{\sigma^2} \int_0^s f'\left(\frac{\sigma^2}{4} L_r(H_r) \right) dr + \frac{2}{\sigma} B_s + \frac{1}{2} L_s(0), \tag{8.1}$$

where $L_s(t)$ denotes the local time of the process H accumulated at level t up to time s. This SDE has a unique weak solution whose law is $\widetilde{\mathbb{P}}^a$ until the first time T^a when it reaches level a. Hence (8.1) has unique weak solution until a possible explosion time $\sup_{a>0} T^a$.

In the (sub)critical case, i.e., when $\Lambda(f) = +\infty$, see Proposition 16, H_s does not explode and returns to zero after any time, $\tau_x = \inf\left\{ s > 0,\ L_s(0) > \frac{4}{\sigma^2} x \right\} < \infty$ a.s., and

$$\{L_{\tau_x}(t),\ t \ge 0, x > 0\} \overset{(d)}{=} \{\mathscr{L}_x(t),\ t \ge 0,\ x > 0\}.$$

In this case, Theorem 14 becomes

Corollary 7. *Assume that f satisfies Assumption (H1'), and moreover that $\Lambda(f) = +\infty$. Then $\{ \frac{\sigma^2}{4} L_{\tau_x}(t),\ t \ge 0, x > 0 \}$ is a weak solution of equation (4.10).*

8.2 The Effect of the Interaction for Large Population

We consider again the $\mathbb{R}_+$-valued two-parameter stochastic process $\{Z^x_t, t \ge 0, x \ge 0\}$ which solves the SDE (4.10), where the function f satisfies Assumption (H1).

Recall Proposition 14 which states that the process $\{Z^x_\cdot, x \ge 0\}$ is a Markov process with values in $C(\mathbb{R}_+, \mathbb{R}_+)$, the space of continuous functions from $\mathbb{R}_+$ into $\mathbb{R}_+$, starting from 0 at $x = 0$. Moreover, we have that whenever $0 < x \le y$, $Z^y_t \ge Z^x_t$ for all $t \geq 0$ a.s. For $x > 0$, define T^x the extinction time of the process Z^x (it is also the height of the forest of the associated genealogical trees) by

$$T^x = \inf\{t > 0, Z_t^x = 0\},$$

and S^x the total mass of Z^x by

$$S^x = \int_0^{T^x} Z_t^x dt.$$

We next study the limits of T^x and S^x as $x \to \infty$. We first need to recall some preliminary results on a class of one-dimensional Kolmogorov diffusions (drifted Brownian motions).

8.2.1 Preliminary Results

Consider a one-dimensional drifted Brownian motion with values in $[0, \infty)$ which is killed when it first hits zero

$$dX_t = q(X_t)dt + dB_t, \qquad X_0 = x > 0,$$

where q is defined and is C^1 on $(0, \infty)$, and $\{B_t, t \geq 0\}$ is a standard one-dimensional Brownian motion. In particular, q is allowed to explode at the origin. We shall assume that

Assumption (H2) There exists $x_0 > 0$ such that $q(x) < 0$, for all $x \geq x_0$, and $\limsup_{x \to 0^+} q(x) < \infty$.

The condition (H2) implies that q is bounded from above by some constant. It ensures that ∞ is inaccessible, in the sense that a.s. ∞ cannot be reached in finite time from $X_0 = x \in (0, \infty)$.

We denote T_y^x the first time the process X hits $y \in [0, \infty)$ when starting from $X_0 = x$

$$T_y^x = \inf\{t > 0 : X_t = y \mid X_0 = x\}.$$

We say that ∞ is an entrance boundary for X (see, for instance, Revuz and Yor [41], page 305) if there exist $y > 0$ and a time $t > 0$ such that

$$\lim_{x \uparrow \infty} \mathbb{P}(T_y^x < t) > 0.$$

Let us introduce the following condition where $Q(y) = 2 \int_1^y q(x)dx$, $y \geq 1$.

Assumption (H3)

$$\int_1^\infty e^{-Q(y)} \int_y^\infty e^{Q(z)} dz dy < \infty.$$

Tonelli's theorem ensures that (H3) is equivalent to

$$\int_1^\infty e^{Q(y)} \int_1^y e^{-Q(z)}\, dz\, dy < \infty.$$

We have the following result, which is Proposition 7.6 in [14].

Proposition 30. *The following are equivalent:*

1) ∞ *is an entrance boundary for* X.
2) *(H3) holds.*
3) *For any* $a > 0$, *there exists* $y_a > 0$ *such that*

$$\sup_{x > y_a} \mathbb{E}\left(e^{a T_{y_a}^x}\right) < \infty.$$

We can now establish

Proposition 31. *Assume that* $(H2)$ *holds. We have*

1) *If* $(H3)$ *does not hold, then for all* $y \geq 0$,

$$\sup_{x > y} T_y^x = \infty \qquad a.s.$$

2) *If* $(H3)$ *holds, then for all* $y \geq 0$,

$$\sup_{x > y} T_y^x < \infty \qquad a.s.,$$

and, moreover, there exists some positive constant c *such that*

$$\sup_{x > 0} \mathbb{E}\left(e^{c T_0^x}\right) < \infty.$$

PROOF:

1) If (H3) does not hold, then by Proposition 30, ∞ is not an entrance boundary for X. It means that for all $y > 0, t > 0$,

$$\lim_{x \uparrow \infty} \mathbb{P}(T_y^x < t) = 0.$$

Hence for all $t > 0$, since $x \to T_y^x$ is increasing a.s.,

$$\mathbb{P}(\sup_{x > y} T_y^x < t) = 0,$$

hence

$$\sup_{x > y} T_y^x = \infty \qquad a.s.$$

2) The result is a consequence of Proposition 30. We can prove it by using the same argument as used in the proof of Theorem 10. $\qquad\square$

It is not obvious when (H3) holds. But from the following result, if q satisfies some explicit conditions, we can decide whether (H3) holds or not.

Proposition 32. *Suppose that* (H2) *holds. We have*

1) *If*

$$\int_{x_0}^{\infty} \frac{1}{q(x)} dx = -\infty \qquad and \qquad \limsup_{x \to \infty} \frac{q'(x)}{q(x)^2} < \infty,$$

then (H3) *does not hold.*

2) *If there exists $q_0 < 0$ such that $q(x) \le q_0$ for all $x \ge x_0$,*

$$\int_{x_0}^{\infty} \frac{1}{q(x)} dx > -\infty \qquad and \qquad \liminf_{x \to \infty} \frac{q'(x)}{q(x)^2} > -2,$$

then (H3) *holds.*

3) *If*

$$\int_{x_0}^{\infty} \frac{1}{q(x)} dx > -\infty \qquad and \qquad q'(x) \le 0 \qquad \forall x \ge x_0,$$

then (H3) *holds.*

PROOF:

1) Define $s(y) := \int_y^{\infty} e^{Q(z)} dz$. If $s(x_0) = \infty$, then $s(y) = \infty$ for all $y \ge x_0$, so that (H3) does not hold.

We consider the case $s(x_0) < \infty$. Integrating by parts on $\int s e^{-Q} dy$ gives

$$\int_{x_0}^{\infty} s e^{-Q} dy = \int_{x_0}^{\infty} \frac{s}{2q} e^{-Q} 2q \, dy = \frac{-s}{2q} e^{-Q} \bigg|_{x_0}^{\infty} - \int_{x_0}^{\infty} \frac{1}{2q} dy - \int_{x_0}^{\infty} s e^{-Q} \frac{q'}{2q^2} dy \tag{8.2}$$

From $\int_{x_0}^{\infty} \frac{1}{q(x)} dx = -\infty$, (8.2) implies that

$$\int_{x_0}^{\infty} s e^{-Q} \left(1 + \frac{q'}{2q^2}\right) dy = \infty.$$

Since $\limsup_{x \to \infty} \frac{q'(x)}{q(x)^2} < \infty$, then $\int_{x_0}^{\infty} s e^{-Q} dy = \infty$. Condition (H3) does not hold.

2) We can easily deduce from $q(x) \le q_0$ for all $x \ge x_0$ that $s(y)$ tends to zero as y tends to infinity, and $s(y)e^{-Q(y)}$ is bounded in $y \ge x_0$. Because $\int_{x_0}^{\infty} \frac{1}{q(x)} dx > -\infty$, (8.2) implies that $s e^{-Q}(1 + \frac{q'}{2q^2})$ is integrable. Then thanks to the condition $\liminf_{x \to \infty} \frac{q'(x)}{q(x)^2} > -2$, we conclude that (H3) holds.

3) From $q(x) \le q(x_0) < 0$ for all $x \ge x_0$, we can easily deduce that $Q(y) \to -\infty$ and $s(y) \to 0$ as $y \to \infty$. Applying the Cauchy's mean value theorem to $s(y)$ and $q_1(y) := e^{Q(y)}$, we have for all $y \ge x_0$, there exists $\xi \in (y, \infty)$ such that

$$\frac{\int_y^{\infty} e^{Q(z)} dz}{e^{Q(y)}} = \frac{s'(\xi)}{q_1'(\xi)} = -\frac{1}{2q(\xi)}.$$

Because $q'(x) \leq 0$ for all $x \geq x_0$, we obtain

$$s(y)e^{-Q(y)} \leq -\frac{1}{2q(y)}, \qquad \text{for all} \qquad y \geq x_0.$$

Hence

$$\int_{x_0}^{\infty} s(y)e^{-Q(y)}dy \leq -\int_{x_0}^{\infty} \frac{1}{2q(y)}dy < \infty.$$

Then (H3) holds.   □

8.2.2 Height of the Continuous Forest of Trees

We consider the process $\{Z_t^x, t \geq 0\}$ solution of (4.10). It follows from the Itô formula that the process $Y_t^x = \sqrt{Z_t^x}$ solves the SDE

$$dY_t^x = \frac{f((Y_t^x)^2) - \sigma^2/4}{2Y_t^x}dt + \frac{\sigma}{2}dW_t, \qquad Y_0^x = \sqrt{x}. \tag{8.3}$$

Note that, the height of the process Z^x is

$$T^x = \inf\{t > 0, Z_t^x = 0\} = \inf\{t > 0, Y_t^x = 0\}.$$

We now establish the large x behaviour of T^x.

Theorem 15. *Assume that f is a function satisfying (H1) and that there exists $a_0 > 0$ such that $f(x) \neq 0$ for all $x \geq a_0$. If $\int_{a_0}^{\infty} \frac{1}{|f(x)|}dx = \infty$, then*

$$T^x \to \infty \quad a.s. \quad as \quad x \to \infty.$$

PROOF: Let β' be a constant such that $\beta' > \beta$. By a well-known comparison theorem, $Y_t^x \geq Y_t^{1,x}$, where $Y_t^{1,x}$ solves

$$dY_t^{1,x} = -\frac{\beta'(Y_t^{1,x})^2 - f((Y_t^{1,x})^2) + \sigma^2/4}{2Y_t^{1,x}}dt + \frac{\sigma}{2}dW_t, \qquad Y_0^{1,x} = \sqrt{x},$$

Note that the function $\beta'x - f(x) + \sigma^2/4$ is positive and increasing, then $f_1(x) := -\frac{\beta'x^2 - f(x^2) + \sigma^2/4}{2x}$ satisfies (H2), and

$$\limsup_{x \to \infty} \frac{f_1'(x)}{f_1(x)^2} < \infty.$$

Moreover from Lemma 16 there exists $x_1 > 0$ such that $\beta'x - f(x) \geq 1$ for all $x \geq x_1$, hence

$$\int_1^\infty \frac{1}{f_1(x)}dx = -\int_1^\infty \frac{2x}{\beta'x^2 - f(x^2) + \sigma^2/4}dx$$

$$= -\int_1^\infty \frac{1}{\beta'x - f(x) + \sigma^2/4}dx$$

$$\leq -\int_1^{x_1} \frac{1}{\beta'x - f(x) + \sigma^2/4}dx - 2\int_{x_1}^\infty \frac{1}{\beta'x - f(x)}dx$$

$$= -\infty,$$

again by Lemma 16. The result now follows readily from Propositions 31 and 32. $\square$

Theorem 16. *Assume that f is a function satisfying (H1) and that there exists $a_0 > 0$ such that $f(x) \neq 0$ for all $x \geq a_0$. If $\int_{a_0}^\infty \frac{1}{|f(x)|}dx < \infty$, then*

$$\sup_{x>0} T^x < \infty \qquad a.s.,$$

and, moreover, there exists some positive constant c such that

$$\sup_{x>0} \mathbb{E}\left(e^{cT^x}\right) < \infty.$$

PROOF: We can rewrite the SDE (8.3) as (with again $\beta' > \beta$)

$$dY_t^x = \frac{\beta'(Y_t^x)^2 - h((Y_t^x)^2)}{2Y_t^x}dt + \frac{\sigma}{2}dW_t, \qquad Y_0^x = \sqrt{x},$$

where $h(x) := \beta'x - f(x) + 1$ is a positive and increasing function. By Lemma 16, we have $\int_1^\infty \frac{1}{h(x)}dx < \infty$ which is equivalent to $\sum_{n=1}^\infty \frac{1}{h(n)} < \infty$. Let

$$a_1 = h(1), \qquad a_n = \min\{h(n), 2a_{n-1}\} \quad \forall n > 1.$$

It is easy to see that for all $n > 1$,

$$a_{n-1} < a_n \leq h(n), \qquad \frac{a_n}{a_{n-1}} \leq 2.$$

We also have

$$\frac{1}{a_1} = \frac{1}{h(1)}$$

$$\frac{1}{a_2} \leq \frac{1}{h(2)} + \frac{1}{2a_1} = \frac{1}{h(2)} + \frac{1}{2h(1)}$$

$$\frac{1}{a_3} \leq \frac{1}{h(3)} + \frac{1}{2a_2} \leq \frac{1}{h(3)} + \frac{1}{2h(2)} + \frac{1}{4h(1)}$$

$$\frac{1}{a_n} \leq \frac{1}{h(n)} + \frac{1}{2a_{n-1}} \leq \frac{1}{h(n)} + \frac{1}{2h(n-1)} + \cdots + \frac{1}{2^{n-1}h(1)}.$$

Therefore

$$\sum_{n=1}^{\infty} \frac{1}{a_n} \leq 2 \sum_{n=1}^{\infty} \frac{1}{h(n)} < \infty.$$

Now, we define a continuous increasing function g as follows. Let h_1 denote a piecewise linear function, which is such that $h_1(n) = a_n$, $n \geq 1$. Define the function h_2 as follows.

$$h_2(x) = \begin{cases} h(x), & 0 \leq x \leq 1 \\ h_1(x), & x \geq 1. \end{cases}$$

We then smoothen all the nodal points of the graph of h_2 to obtain a smooth curve which is the graph of an increasing function g_1. Let $g(x) = \frac{1}{2}g_1(x)$. We have for all $n \geq 1$ and $x \in [n, n+1)$,

$$h(x) \geq h(n) \geq a_n \geq \frac{1}{2}a_{n+1} = g(n+1) \geq g(x).$$

By the comparison theorem, $Y_t^x \leq Y_t^{2,x}$, where $Y_y^{2,x}$ solves

$$dY_t^{2,x} = \frac{\beta'(Y_t^{2,x})^2 - g((Y_t^{2,x})^2)}{2Y_t^{2,x}} dt + \frac{\sigma}{2} dW_t, \qquad Y_0^{2,x} = \sqrt{x}.$$

Since

$$\sum_{n=1}^{\infty} \frac{1}{g(n)} = 2 \sum_{n=1}^{\infty} \frac{1}{a_n} < \infty,$$

we deduce that $\int_1^{\infty} \frac{1}{g(x)} dx < \infty$, and $\frac{g(x)}{x} \to \infty$ as $x \to \infty$, by Lemma 16. Let $f_2(x) := \frac{\beta'x^2 - g(x^2)}{2x}$, then there exist $x_1 > 0, q_1 < 0$ such that $f_2(x) < q_1$ for all $x \geq x_1$, and

$$\int_{x_1}^{\infty} \frac{1}{f_2(x)} dx = \int_{x_1}^{\infty} \frac{2x}{\beta'x^2 - g(x^2)} dx = \int_{x_1^2}^{\infty} \frac{1}{\beta'x - g(x)} dx > -\infty.$$

Moreover,

$$\liminf_{x \to \infty} \frac{f_2'(x)}{f_2(x)^2} = \liminf_{x \to \infty} \frac{-4xg'(x)}{g(x)^2}.$$

But for all $x \in [n, n+1)$,

$$\frac{g'(x)x}{g(x)^2} \leq \frac{(n+1)}{g(n)^2} \max_{i \in \{n-1,n,n+1\}} \{g(i+1) - g(i)\} < \frac{(n+1)g(n+2)}{g(n)^2} \leq \frac{4(n+1)}{g(n)} \to 0,$$

as $n \to \infty$. The result follows from Propositions 31 and 32. $\square$

8.2.3 Total Mass of the Continuous Forest of Trees

Recall that in the continuous case, the total mass of the genealogical tree is given as

$$S^x = \int_0^{T^x} Z_t^x \, dt$$

Consider the increasing process

$$A_t^x = \int_0^t Z_s^x \, ds, t \geq 0,$$

and the associated time change

$$\eta^x(t) = \inf\{s > 0, A_s > t\}.$$

We now define $U_t^x = \frac{1}{\sigma} Z^x \circ \eta^x(t), t \geq 0$. It is easily seen that the process U^x solves the SDE

$$dU_t^x = \frac{f(\sigma U_t^x)}{\sigma^2 U_t^x} dt + dW_t, \qquad U_0^x = \frac{x}{\sigma}. \tag{8.4}$$

Let $\tau^x := \inf\{t > 0, U_t^x = 0\}$. It follows from above that $\eta^x(\tau^x) = T^x$, hence $S^x = \tau^x$. We have

Theorem 17. *Suppose that the function $\frac{f(x)}{x}$ satisfies (H1) and there exists $a_0 > 0$ such that $f(x) \neq 0$ for all $x \geq a_0$.*

1) If $\int_{a_0}^{\infty} \frac{x}{|f(x)|} dx = \infty$, then

$$S^x \to \infty \quad a.s. \quad as \quad x \to \infty.$$

2) If $\int_{a_0}^{\infty} \frac{x}{|f(x)|} dx < \infty$, then

$$\sup_{x>0} S^x < \infty \qquad a.s.,$$

and, moreover, there exists some positive constant c such that

$$\sup_{x>0} \mathbb{E}\left(e^{cS^x}\right) < \infty.$$

PROOF: Note that we can rewrite the SDE (8.4) as

$$dU_t^x = \left(\beta' U_t^x - h(U_t^x)\right) dt + dW_t, \qquad U_0^x = \frac{x}{\sigma},$$

where $h(x) := \beta' x - \frac{f(\sigma x)}{\sigma^2 x}$, with again $\beta' > \beta$, is a positive and increasing function.

1) By the comparison theorem, $U_t^x \geq U_t^{1,x}$, where $U_t^{1,x}$ solves

$$dU_t^{1,x} = -h(U_t^{1,x})dt + dW_t, \qquad U_0^{1,x} = \frac{x}{\sigma}.$$

The result follows from Proposition 31, Proposition 32, and Lemma 16.
2) The result is a consequence of Propositions 31 and 32. We can prove it by using the same argument as used in the proof of Theorem 16.

$\square$

Example 2. As in the case of a finite population, let $f(z) = -ae^{bx}$, with $a, b > 0$. If $b \leq 1$, then $T^x \to \infty$ and $S^x \to \infty$, as $x \to \infty$. If $1 < b \leq 2$, then there exists $c > 0$ such that $\sup_{x \geq 1} \mathbb{E}\left[e^{cT^x}\right] < \infty$, while $S^x \to \infty$ as $x \to \infty$. Finally if $b > 2$, then for some $c > 0$, both $\sup_{x \geq 1} \mathbb{E}\left[e^{cT^x}\right] < \infty$ and $\sup_{x \geq 1} \mathbb{E}\left[e^{cS^x}\right] < \infty$.

Appendix

The aim of this Appendix is to describe most of the technical results which we are using in all the previous chapters. Section A.1 presents some of the results from stochastic calculus which we are using, with the main emphasis on the calculus for discontinuous processes, which is less known. Section A.2 states a fundamental martingale representation theorem of continuous martingales as stochastic integrals with respect to Brownian motion. Section A.3 describes the connection, initially due to Stroock and Varadhan, see [42], between martingale problems and SDEs. The reason why this connection is essential is that when taking a weak limit of a sequence of approximate models, the limit is naturally formulated as the solution of a martingale problem, i.e., a weak solution of an SDE, as we explain here. Section A.4 gives the definition of the local time of a continuous semimartingale and derives the occupation times formula, which plays an important role in the previous chapters. We also express reflected Brownian motion in terms of its local time at the levels of reflection. Section A.5 states two Girsanov theorems which are used in Chapter 7, namely the one for Brownian motion and the one for point processes. Section A.6 states the celebrated Lévy–Khintchine formula. Finally section A.7 presents some important facts about tightness and weak convergence of (possibly discontinuous) processes. In this chapter, only few proofs are given. Instead, we give precise references to the literature.

A.1 Some Elements of Stochastic Calculus

We give here only a short overview of the results which we are using in these Notes. We refer the reader among many possible references to Protter [40] for a complete presentation. Since we treat almost only scalar-valued processes, we present the necessary basic facts from stochastic calculus only in the scalar case. In this section, we suppose given a probability space $(\Omega, \mathscr{F}, \mathscr{F}_t, \mathbb{P})$ equipped with the filtration $\{\mathscr{F}_t, t \geq 0\}$ (i.e., the $\mathscr{F}_t$'s are sub-σ-algebras of $\mathscr{F}$, and $\mathscr{F}_s \subset \mathscr{F}_t$ whenever $s \leq t$). Each $\mathscr{F}_t$ is assumed to contain all the $\mathbb{P}$-null sets of $\mathscr{F}$. We recall

© Springer International Publishing Switzerland 2016
É. Pardoux, *Probabilistic Models of Population Evolution*, Mathematical
Biosciences Institute Lecture Series 1.6, DOI 10.1007/978-3-319-30328-4

Definition 2. The sigma-algebra of progressively measurable subsets of $\Omega \times [0,+\infty)$ is the class of subsets $A \subset \Omega \times [0,+\infty)$ such that for all $t \geq 0$,

$$A \cap (\Omega \times [0,t]) \in \mathscr{F}_t \otimes \mathscr{B}_{[0,t]},$$

where $\mathscr{B}_{[0,t]}$ denotes the sigma-algebra of Borel subsets of $[0,t]$.

The sigma-algebra of predictable subsets of $\Omega \times [0,+\infty)$ is the smallest σ-algebra which contains all the sets of the form $A_s \times (s,t]$, where $0 \leq s < t$ and $A_s \in \mathscr{F}_s$.

All processes below will be supposed to be progressively measurable. A progressively measurable process with left-continuous trajectories is predictable.

Definition 3. A martingale is a process $\{M_t,\ t \geq 0\}$ which satisfies

(i) M_t is $\mathscr{F}_t$-measurable and integrable for all $t \geq 0$ and
(ii) $\mathbb{E}[M_t|\mathscr{F}_s] = M_s$, whenever $s < t$.

A local martingale is a process $\{M_t,\ t \geq 0\}$ to which we can associate a sequence $\{T_n,\ n \geq 1\}$ of stopping times such that

(i) $T_n \uparrow +\infty$, as $n \to \infty$ and
(ii) For each $n \geq 1$, $\{M_t^{T_n} := M_{t \wedge T_n},\ t \geq 0\}$ is a martingale.

In this chapter, we use stochastic calculus for two distinct classes of semimartingales.

The first class is the class of continuous semimartingales (sum of a local martingale and a process of bounded variation on any finite interval), whose martingale part is a stochastic integral with respect to Brownian motion. More precisely, let $(\Omega, \mathscr{F}, \mathscr{F}_t, \mathbb{P})$ be a probability space with the filtration $\{\mathscr{F}_t,\ t \geq 0\}$, and $\{B_t,\ t \geq 0\}$ be an $\mathscr{F}_t$-Brownian motion, that is, a continuous $\mathscr{F}_t$-martingale, which is such that $B_0 = 0$ and $B_t^2 - t$ is also an $\mathscr{F}_t$-martingale (see Theorem 18 below). Suppose now that $\{\psi_t, \varphi_t,\ t \geq 0\}$ are $\mathscr{F}_t$-progressively measurable processes (this means that for any $t > 0$, $(\omega, s) \to (\psi(\omega, s), \varphi(\omega, s))$ is $\mathscr{F}_t \otimes \mathscr{B}([0,t])$ measurable from $\Omega \times [0,t]$ into $\mathbb{R}^d \times \mathbb{R}^{d \times k}$), such that for any $T > 0$,

$$\int_0^T [|\psi_t| + |\varphi_t|^2] dt < \infty \quad \text{a. s.,}$$

X_0 is an $\mathscr{F}_0$-measurable random variable, and

$$X_t = X_0 + \int_0^t \psi_s ds + \int_0^t \varphi_s dB_s, \quad t \text{ a. s.}$$

Then we have the Itô formula : for any $f \in C^2(\mathbb{R})$, $t \geq 0$,

$$f(X_t) = f(X_0) + \int_0^t \left[f'(X_s)\psi_s + \frac{1}{2}f''(X_s)\varphi_s^2 \right] ds + \int_0^t f'(X_s)\varphi_s dB_s.$$

In the case where the process $\int_0^t \psi_s ds$ is replaced by a more general continuous finite variation (denoted below FV) process V_t, the Itô formula reads

$$f(X_t) = f(X_0) + \int_0^t f'(X_s) dV_s + \frac{1}{2} \int_0^t f''(X_s) \varphi_s^2 ds + \int_0^t f'(X_s) \varphi_s dB_s,$$

where the first integral on the right is a Stieltjes integral. Let us write $M_t = \int_0^t \varphi_s dB_s$ for the local martingale part of X_t. We have that

$$\langle M \rangle_t = [M]_t = \int_0^t \varphi_s^2 ds,$$

where the quadratic variation $[M]$ of the continuous or discontinuous martingale M is defined as

$$[M]_t = M_t^2 - 2 \int_0^t M_{s^-} dM_s \qquad (A.1)$$

and the conditional quadratic variation $\langle M \rangle$ of M is the unique predictable process such that $[M]_t - \langle M \rangle_t$ is a martingale.

Let us now write Itô's formula in the case of a continuous semimartingale of the form

$$X_t = X_0 + V_t + M_t,$$

where $\{V_t\}$ is a finite variation continuous process and $\{M_t\}$ is a continuous local martingale. If $f \in C^2(\mathbb{R})$,

$$f(X_t) = f(X_0) + \int_0^t f'(X_s) dV_s + \int_0^t f'(X_s) dM_s + \frac{1}{2} \int_0^t f''(X_s) d\langle M \rangle_s. \qquad (A.2)$$

The second class of semimartingales which we need to use here is a class of discontinuous finite variation semimartingales. If $\{X_t, \ t \geq 0\}$ is a finite variation right-continuous $\mathbb{R}$-valued process and $f \in C^1(\mathbb{R})$,

$$f(X_t) = f(X_0) + \int_0^t f'(X_{s^-}) dX_s + \sum_{0 \leq s \leq t} \left[f(X_s) - f(X_{s^-}) - f'(X_{s^-}) \Delta X_s \right],$$

where $\Delta X_s = X_s - X_{s^-}$, and the above sum is over those s such that $\Delta X_s \neq 0$. The above formula follows by considering both the evolution of $f(X_s)$ between the jumps of X_s and the jumps of $f(X_s)$ produced by those of X_s. Note that in the case $f(x) = x^2$, the above formula reduces to

$$(X_t)^2 = (X_0)^2 + 2 \int_0^t X_{s^-} dX_s + \sum_{0 \leq s \leq t} (\Delta X_s)^2. \qquad (A.3)$$

If X is the sum of a continuous FV process and an FV martingale $\{M_t\}$, then

$$\sum_{0 \leq s \leq t} (\Delta X_s)^2 = \sum_{0 \leq s \leq t} (\Delta M_s)^2 = [M]_t. \qquad (A.4)$$

The last identity follows by comparing (A.1) and (A.3).

A.2 A Martingale Representation Theorem

We state and prove the results in dimension 1, since we need them only in that case. We first establish

Theorem 18. *(Paul Lévy) Let $\{M_t,\ t \geq 0\}$ be a continuous martingale such that $M_0 = 0$ and whose quadratic variation satisfies $\langle M \rangle_t = t$, for all $t \geq 0$. Then $\{M_t,\ t \geq 0\}$ is a standard Brownian motion.*

PROOF: It follows from Itô's formula for the continuous martingale M_t that $N_t := \exp(iuM_t + u^2t/2)$ is a martingale. Consequently for $0 \leq s < t$, $\mathbb{E}^{\mathscr{F}_s}N_t = N_s$, which implies that

$$\mathbb{E}^{\mathscr{F}_s} \exp[iu(M_t - M_s)] = \exp(-u^2(t-s)/2),$$

hence $M_t - M_s$ is an $\mathscr{N}(0, t-s)$ random variable, which is independent of $\{M_r,\ 0 \leq r \leq s\}$. The result follows.   □

We now prove the martingale representation theorem.

Theorem 19. *Let $\{M_t,\ t \geq 0\}$ be a one-dimensional continuous martingale, defined on a probability space $(\Omega, \mathscr{F}, \mathscr{F}_t, \mathbb{P})$, with associated increasing process*

$$\langle M \rangle_t = \int_0^t A_s ds,$$

where $\{A_t,\ t \geq 0\}$ is an $\mathscr{F}_t$-progressively measurable $\mathbb{R}_+$-valued process. Then there exists, possibly on an enlarged probability space $(\Omega', \mathscr{F}', \mathbb{P}')$, a standard Brownian motion $\{B_t\}$ such that

$$M_t = \int_0^t \sqrt{A_s} dB_s,\ t \geq 0. \tag{A.5}$$

PROOF: Let $(\Omega', \mathscr{F}', \mathbb{P}') = (\Omega \times C(\mathbb{R}_+), \mathscr{F} \otimes \mathscr{C}, \mathbb{P} \times \mathscr{W})$, where $\mathscr{C}$ denotes the Borel σ field over $C(\mathbb{R}_+)$, and $\mathscr{W}$ the Wiener measure on $(C(\mathbb{R}_+), \mathscr{C})$. Let $\{W_t,\ t \geq 0\}$ denote the canonical process defined on $(C(\mathbb{R}_+), \mathscr{C}, \mathscr{W})$. Define the two following $\mathscr{F}_t$-progressively measurable processes

$$\text{For } t \geq 0, \quad a_t = \begin{cases} A_t^{-1/2}, & \text{if } A_t > 0, \\ 0, & \text{if } A_t = 0. \end{cases}$$

$$b_t = \begin{cases} 0, & \text{if } A_t > 0, \\ 1, & \text{if } A_t = 0. \end{cases}$$

Let now

$$B_t = \int_0^t a_s dM_s + \int_0^t b_s dW_s,\ t \geq 0.$$

It is easy to check that $\{B_t,\ t \geq 0\}$ is a continuous martingale. Moreover, since M and W are independent,

$$\langle B \rangle_t = \int_0^t a_s^2 A_s ds + \int_0^t b_s^2 ds$$

$$= \int_0^t \mathbf{1}_{\{A_s > 0\}} ds + \int_0^t \mathbf{1}_{\{A_s = 0\}} ds$$

$$= t,$$

hence from Lévy's Theorem 18, $\{B_t, \ t \geq 0\}$ is a Brownian motion. It remains to establish (A.5). It is plain that

$$\int_0^t \sqrt{A_s} dB_s = \int_0^t \mathbf{1}_{\{A_s > 0\}} dM_s.$$

Now

$$N_t = M_t - \int_0^t \mathbf{1}_{\{A_s > 0\}} dM_s$$

is a martingale, and

$$\langle N \rangle_t = \int_0^t A_s \mathbf{1}_{\{A_s = 0\}} ds = 0,$$

hence the result. $\qquad \qquad \square$

A.3 Martingale Problems and SDEs

Let $\{X_t, \ t \geq 0\}$ be a continuous progressively measurable process defined on a probability space $(\Omega, \mathscr{F}, \mathscr{F}_t, \mathbb{P})$, such that, for some Borel measurable and locally bounded functions $b, \sigma : \mathbb{R} \to \mathbb{R}$, all $t \geq 0$, and some continuous local martingale M_t,

$$X_t = X_0 + \int_0^t b(X_s) ds + M_t, \quad \text{where}$$

$$\langle M \rangle_t = \int_0^t \sigma^2(X_s) ds.$$

It then follows from Theorem 19 that (possibly on a larger probability space) there exists a Brownian motion $\{B_t\}$ such that

$$X_t = X_0 + \int_0^t b(X_s) ds + \int_0^t \sigma(X_s) dB_s.$$

In such a case one says that X is a weak solution of this last SDE (a strong solution is associated to an a priori given Brownian motion).

Of course, the same result is true under the following stronger assumption: for any $f \in C^2(\mathbb{R})$, the process

$$M_t^f = f(X_t) - f(X_0) - \int_0^t Af(X_s) ds$$

is a local martingale, where

$$Af(x) := b(x)f'(x) + \frac{1}{2}\sigma^2(x)f''(x).$$

A.4 Local Time and the Occupation Times Formula

In this section, we are given a continuous semimartingale X_t of the form

$$X_t = V_t + B_t,$$

where V_t is a continuous bounded variation process, and B_t is a standard Brownian motion. We state the

Definition 4. For any $a \in \mathbb{R}$, we define the local time of the semimartingale X at level a as the increasing process given by Tanaka's formula

$$L_t^a = 2(X_t - a)^+ - 2(X_0 - a)^+ - 2\int_0^t \mathbf{1}_{\{X_s > a\}} dX_s.$$

One can choose a modification of L_t^a which is continuous in t and right-continuous in a.

Proposition 33. *Occupation times formula for Brownian motion.*

$$\int_0^t f(X_s)ds = \int_{\mathbb{R}} f(a)L_t^a da$$

$\mathbb{P}$-a.s. for every $t > 0$ and f Borel from $\mathbb{R}$ into $\mathbb{R}_+$.

PROOF: If $f = h''$ with $h \in C^2(\mathbb{R})$, the formula can be established by comparing the definition of L_t^a and Itô's formula. It is then not very hard to show that the formula holds simultaneously for every $t > 0$ and $f \in C^0(\mathbb{R})$ outside a negligible set. An application of the monotone class theorem ends the proof. We refer the reader to the proof of Corollary VI.1.6 of [41] for more details. □

The definition of the local time would be the same if we replace the martingale part of X_t by a continuous martingale different from a standard Brownian motion, but the occupation times formula would be different. Suppose that

$$X_t = V_t + \int_0^t \varphi_t dB_t,$$

where φ_t is progressively measurable and $\int_0^t |\varphi_s|^2 ds < \infty$ a.s. for all $t > 0$. In this case, the occupation times formula becomes

Proposition 34. *Occupation times formula for continuous semimartingale.*

$$\int_0^t f(X_s)\varphi_s^2 ds = \int_{\mathbb{R}} f(a)L_t^a da.$$

$\mathbb{P}$-*a.s. for every* $t > 0$ *and* f *Borel from* $\mathbb{R}$ *into* $\mathbb{R}_+$.

Let now X_t be Brownian motion with drift reflected above 0, i.e., with $\gamma, a \in \mathbb{R}$,

$$X_t = X_0 + \gamma t + aB_t + K_t, \quad X_t \geq 0$$

K_t is continuous and increasing, $K_t = \int_0^t \mathbf{1}_{\{X_s=0\}} dK_s$.

It is easy to show that $K_t = -\inf_{0 \leq s \leq t}([X_0 + \gamma s + aB_s] \wedge 0)$. Since $X_t \geq 0$, from the definition of the local time of X_t at level 0,

$$\frac{1}{2}L_t^0 = X_t - X_0 - \int_0^t \mathbf{1}_{\{X_s>0\}} dX_s.$$

But

$$\int_0^t \mathbf{1}_{\{X_s>0\}} dK_s = 0, \text{ while } \int_0^t \mathbf{1}_{\{X_s>0\}} dB_s = B_t,$$

since a.s., $\mathbf{1}_{\{X_s>0\}} = 1$ ds-a.e. Indeed it follows from Proposition 34 that $\int_0^t \mathbf{1}_{\{X_s=0\}} ds = 0$. Consequently $\frac{1}{2}L_t^0 = K_t$.

Similarly Brownian motion reflected in the interval $[0, a]$ reads

$$X_t = X_0 + \gamma t + aB_t + \frac{1}{2}L_t^0 - \frac{1}{2}L_t^{a^-},$$

where L_t^0 is the local time of X_t at level and $L_t^{a^-}$ is the limit as $\varepsilon \to 0$ of $L_t^{a-\varepsilon}$. Note that since $X_s \leq a$ for all $0 \leq s \leq t$, and $x \to L_t^x$ is right-continuous, $L_t^a = 0$.

The above formulas would be the same, had we replaced $B_t + \gamma t$ by a more general continuous semimartingale.

A.5 Two Girsanov Theorems

We state two versions of the Girsanov theorem, one for the Brownian and one for the point process case. The first one can be found, e.g., in [35] and the second one combines Theorems T2 and T3 from [12], pages 165–166. We assume here that our probability space $(\Omega, \mathbb{P}, \mathscr{F})$ is such that $\mathscr{F} = \sigma(\cup_{t>0}\mathscr{F}_t)$.

Proposition 35. *Let* $\{B_s, s \geq 0\}$ *be a standard d-dimensional Brownian motion (i.e., its coordinates are mutually independent standard scalar Brownian motions)* *defined on the filtered probability space* $(\Omega, \mathscr{F}, \mathbb{P})$. *Let moreover* ϕ *be an* $\mathscr{F}$-*progressively measurable d-dimensional process satisfying* $\int_0^s |\phi(r)|^2 dr < \infty$ *for all* $s \geq 0$. *Let*

$$Y_s = \exp\left\{ \int_0^s \langle \phi(r), dB_r \rangle - \frac{1}{2} \int_0^s |\phi(r)|^2 dr \right\}.$$

If $\mathbb{E}[Y_s] = 1$, $s \geq 0$, *then* $\tilde{B}_s := B_s - \int_0^s \phi(r)dr$, $s \geq 0$, *is a standard Brownian motion under the unique probability measure* $\tilde{\mathbb{P}}$ *on* $(\Omega, \mathscr{F})$ *which is such that* $d\tilde{\mathbb{P}}|_{\mathscr{F}_s}/d\mathbb{P}|_{\mathscr{F}_s} = Y_s$, *for all* $s \geq 0$.

We recall that a stopping time τ is a $[0, +\infty]$-valued r.v. which satisfies $\{\tau \leq t\} \in \mathscr{F}_t$, for all $t > 0$. To a stopping time τ we associate the σ-algebra $\mathscr{F}_\tau = \{A \in \mathscr{F}, A \cap \{\tau \leq t\} \in \mathscr{F}_t, \forall t \geq 0\}$. We now establish the

Corollary 8. *Let the assumptions of Proposition 35 be satisfied. Let* τ *be a stopping time, such that* $\mathbb{P}(\tau < \infty) = \tilde{\mathbb{P}}(\tau < \infty) = 1$. *Then* $\tilde{\mathbb{P}}|_{\mathscr{F}_\tau} << \mathbb{P}|_{\mathscr{F}_\tau}$ *and*

$$\frac{d\tilde{\mathbb{P}}|_{\mathscr{F}_\tau}}{d\mathbb{P}|_{\mathscr{F}_\tau}} = Y_\tau.$$

PROOF: Let $A \in \mathscr{F}_\tau$. Then $A \cap \{\tau \leq t\} \in \mathscr{F}_t$. We have

$$\tilde{\mathbb{P}}(A \cap \{\tau \leq t\}) = \int_{A \cap \{\tau \leq t\}} Y_t d\mathbb{P}$$

$$= \int_{A \cap \{\tau \leq t\}} Y_{t \wedge \tau} d\mathbb{P}$$

$$= \int_{A \cap \{\tau \leq t\}} Y_\tau d\mathbb{P},$$

where the second equality follows from the martingale property of Y_t, and the third from the fact that we integrate on a set where $\tau \leq t$. It remains to let $t \to \infty$ with the help of the monotone convergence theorem, and use the assumptions to conclude that

$$\tilde{\mathbb{P}}(A) = \int_A Y_\tau d\mathbb{P}.$$

$\square$

The notions of Poisson point process and Poisson counting process with a given fixed intensity have been introduced in sections 2.2.2 and 5.2. Clearly, if $P(t)$ is a standard (i.e., rate 1) counting Poisson process and $\lambda > 0$, then $P(\lambda t)$ is a rate λ Poisson process. At various places in this document, starting with section 3.4, we have worked with counting processes of the type $Q(t) = P(\int_0^t \lambda(s)ds)$, where P is a standard Poisson process and $\lambda(t)$ is a stochastic process. It is quite natural to call $\lambda(t)$ the intensity of the counting process $Q(t)$. If we identify the counting process $Q(t)$ with the points which it counts, we can as well call it a point process. This explains the terminology in the next statement.

Proposition 36. *Let $\{(Q_s^{(1)},\ldots,Q_s^{(d)}),s \geq 0\}$ be a d-variate point process adapted to some filtration $\mathscr{F}$, and let $\{\lambda_s^{(i)},s \geq 0\}$ be the predictable $(\mathbb{P},\mathscr{F})$-intensity of $Q^{(i)}, 1 \leq i \leq d$. Assume that none of the $Q^{(i)}$, $Q^{(j)}$, $i \neq j$, jump simultaneously. Let $\{\mu_r^{(i)},r \geq 0\},1 \leq i \leq d$, be nonnegative $\mathscr{F}$-predictable processes such that for all $s \geq 0$ and all $1 \leq i \leq d$*

$$\int_0^s \mu_r^{(i)} \lambda_r^{(i)} dr < \infty \qquad \mathbb{P}\text{-a.s.}$$

For $i = 1,\ldots,d$ and $s \geq 0$ define, $\{T_k^i, k = 1,2\ldots\}$ denoting the jump times of $Q^{(i)}$,

$$Y_s^{(i)} = \left(\prod_{k \geq 1 : T_k^i \leq s} \mu_{T_k^i}^{(i)} \right) \exp\left\{ \int_0^s (1 - \mu_r^{(i)})\lambda_r^{(i)} dr \right\} \quad and \quad Y_s = \prod_{j=1}^d Y_s^{(j)}, \quad s \geq 0.$$

If $\mathbb{E}[Y_s] = 1$, $s \geq 0$, then, for each $1 \leq i \leq d$, the process $Q^{(i)}$ has the $(\tilde{\mathbb{P}},\mathscr{F})$-intensity $\tilde{\lambda}_r^{(i)} = \mu_r^{(i)}\lambda_r^{(i)}$, $r \geq 0$, where the probability measure $\tilde{\mathbb{P}}$ is defined by $d\tilde{\mathbb{P}}|_{\mathscr{F}_s}/d\mathbb{P}|_{\mathscr{F}_s} = Y_s, s \geq 0$.

A.6 The Lévy–Khintchine Formula

Definition 5. The law of the real-valued r.v. X is said to be infinitely divisible if for every $n > 1$, there exist n i.i.d. r.v.'s $X_1^n,\ldots,X_n^n$ such that

$$\mathscr{L}(X) = \mathscr{L}(X_1^n + \cdots + X_n^n).$$

The characteristic function of an infinitely divisible r.v. X can be written as

$$\varphi_X(u) = \mathbb{E}[\exp(iuX)] = \exp(-\Psi(u)),$$

with a unique *characteristic exponent* $\Psi \in C(\mathbb{R};\mathbb{C})$ satisfying $\Psi(0) = 0$, specified by the celebrated Lévy–Khintchine formula (see, e.g., [11])

Theorem 20. *A function $\Psi : \mathbb{R} \in \mathbb{C}$ is the characteristic exponent of an infinitely divisible distribution on $\mathbb{R}$ iff there are $\alpha \in \mathbb{R}$, $\beta \geq 0$, Λ a measure on $\mathbb{R}\backslash\{0\}$, called the Lévy measure, which satisfies $\int (1 \wedge |x|^2)\Lambda(dx) < \infty$, such that*

$$\Psi(u) = i\alpha u + \beta u^2 + \int_\mathbb{R} (1 - e^{iux} + iux\mathbf{1}_{\{|x|\leq 1\}})\Lambda(dx). \tag{A.6}$$

In the particular case of a positive valued infinitely divisible r.v., we prefer to describe the Laplace exponent of X, i.e., for $\lambda \geq 0$

$$\psi(\lambda) = -\log\{\mathbb{E}[\exp(-\lambda X)]\}$$

which takes the form

$$\psi(\lambda) = d\lambda + \int_0^\infty (1 - e^{-\lambda x})\Lambda(dx), \qquad (A.7)$$

again with $\lambda \geq 0$. Here $d \geq 0$ is called the drift coefficient, and the Lévy measure Λ is a measure on $(0, +\infty)$ which in the present case satisfies

$$\int_{\mathbb{R}_+} (1 \wedge x)\Lambda(dx) < \infty.$$

A.7 Tightness and Weak Convergence in $D([0, +\infty))$

Let us present a sufficient condition for tightness and a weak convergence criterion, which are used several times in these Notes. We refer the reader to [10] or [18] for detailed treatments of this topic, as well as to [21] for some useful results.

Consider a sequence $\{X_t^n, \, t \geq 0\}_{n \geq 1}$ of one-dimensional semimartingales, which is such that for each $n \geq 1$,

$$X_t^n = X_0^n + \int_0^t \varphi_s^n ds + M_t^n, \; 0 \leq t \leq T;$$

$$\langle M^n \rangle_t = \int_0^t \psi_s^n ds, \; t \geq 0;$$

where for each $n \geq 1$, $M_.^n$ is a locally square integrable martingale, φ^n and ψ^n are progressively measurable processes with value in $\mathbb{R}$ and $\mathbb{R}_+$, respectively. Since our martingales $\{M_t^n, \, t \geq 0\}$ will be discontinuous, we need to consider their trajectories as elements of $D([0, +\infty))$, the space of right-continuous functions with left limits at every point, from $[0, +\infty)$ into $\mathbb{R}$, which we equip with the Skorokhod topology, see Billingsley [10]. Let us first state Aldous's tightness criterion [1]. Recall that a stopping time for the process X^n is a nonnegative r.v. τ which is such that $\{\tau \leq t\} \in \sigma\{X_s^n, \, 0 \leq s \leq t\}$ for all $t > 0$. Consider the condition

(A) $\begin{cases} \text{For each positive } \varepsilon, \, \eta, \, T \text{ there exist } \delta_0 \text{ and } n_0 \text{ such that if } \delta \leq \delta_0, \\ n \geq n_0, \text{and } \tau \text{ is a discrete } X^n\text{- stopping time with } \tau \leq T \text{ a.s.,} \\ \text{then} \qquad \mathbb{P}(|X_{\tau+\delta}^n - X_\tau^n| \geq \varepsilon) \leq \eta. \end{cases}$

Theorem 16.10 in [10] shows that if $\sup_{0 \leq t \leq T} |X_t^n|$ is tight for all $T > 0$ and condition (A) holds, then $(X^n)_{n \geq 1}$ is tight in $D([0, +\infty))$. The following is a consequence of that result, combined with Theorem 13.4 in [10].

Proposition 37. *A sufficient condition for the sequence $\{X_t^n, \, t \geq 0\}_{n \geq 1}$ to be tight in $D([0, \infty))$ is that both*

the sequence of r.v.'s $\{X_0^n, \, n \geq 1\}$ is tight; $\qquad (A.8)$

and for some $c > 0$,

$$\sup_{n \geq 1, s > 0} (|\varphi_s^n| + \psi_s^n) \leq c. \tag{A.9}$$

If, moreover, for any $T > 0$, as $n \to \infty$,

$$\sup_{0 \leq t \leq T} |M_t^n - M_{t-}^n| \to 0 \quad \text{in probability,}$$

then any limit X of a weakly converging subsequence of the original sequence $\{X^n\}_{n \geq 1}$ is a.s. continuous.

Remark 14. Condition (A.9) in the above Proposition can be relaxed. In particular, each of the two following conditions is a sufficient condition which, together with (A.8), implies tightness of $\{X_t^n, t \geq 0\}_{n \geq 1}$ in $D([0, \infty))$.

1. For any $T > 0$, the sequence of r.v.'s

$$\sup_{0 \leq t \leq T} (|\varphi_t^n| + \psi_t^n) \quad \text{is tight.}$$

2. For some $p > 1$ and any $T > 0$, the sequence of r.v.'s

$$\int_0^T \{|\varphi_t^n|^p + \|\psi_t^n\|^p\} dt \quad \text{is tight.}$$

We have moreover

Proposition 38. *Suppose that all conditions of Proposition 37 are satisfied, and that, moreover, as $n \to \infty$,*

$$(X_0^n, \varphi^n \, \psi^n) \Rightarrow (X_0, \varphi, \psi)$$

weakly in $\mathbb{R} \times L_{loc}^1([0, +\infty)) \times L_{loc}^1([0, +\infty))$.

Then X^n converges weakly in $D([0, \infty))$ towards a continuous process X which is such that

$$X_t = X_0 + \int_0^t \varphi_s ds + M_t,$$

where M_t is a continuous local martingale such that

$$\langle M \rangle_t = \int_0^t \psi_s ds.$$

Bibliographical Comments

In chapter 2, section 2.1 follows the treatment in Lyons and Peres [31], while section 2.2 has been inspired by Lambert [23]. Section 3.2 presents results from O'Connell [32]. The material of chapter 4 is mainly taken from Dawson and Li [15]. Section 4.2 is inspired by [23], Bertoin and Le Gall [7], and Le Gall [27]. Section 4.3 presents some results from O'Connell [33]. Section 4.5 is taken from Bertoin, Font-bona, and Martinez [8]. Section 4.6 combines results from [15] and from Le, Pardoux, and Wakolbinger [29]. Chapter 5 presents results from Ba, Pardoux, and Sow [6], see also Le Gall [26] and Pitman and Winkel [39] for results related to section 5.2. Most of chapter 6 presents results from [29] and Ba and Pardoux [4], while section 6.4 is taken from Le and Pardoux [28], see also Ba and Pardoux [5] for the case where f is a power function. Chapter 7 presents results from [4] and a generalization of some of the results in [29]. Section 8.1 generalizes results in [29], Pardoux and Wakolbinger [36], Pardoux and Wakolbinger [37], and [4]. Finally section 8.2 is taken from [28], see also [5] for the case where f is a power function.

All the results concerning the genealogy for populations with interaction are due to the author, jointly with Anton Wakolbinger and with his two former Ph.D. students Mamadou Ba and Vi Le. Proposition 29 and Theorem 14 are new. The Theorem was proved for a logistic (i.e., quadratic) nonlinear function f in [29] via approximation by a discrete population, and for the same class of f's using a stochastic calculus argument in [36]. This last proof was generalized to the case of a general (sub)critical f in [4]. Here we treat the general case, without the (sub)criticality assumption, using the proof via approximation, in a form which has been made concise. The idea of reflecting the contour process in the supercritical case was first introduced by Delmas in the continuous branching case in [16], and has been also used in [6].

References

[1] D. Aldous, Stopping times and tightness. Ann. Probab. **6**, 335–340 (1978)

[2] D. Aldous, The continuum random tree I. Ann. Probab. **19**, 1–28 (1991)

[3] W.J. Anderson, *Continuous-Time Markov Chains* (Springer, New York, 1991)

[4] M. Ba, E. Pardoux, Branching process with competition and a generalized Ray Knight Theorem. Ann. Inst. Henri Poincaré Probab. Stat. **51**, 1290–1313 (2015)

[5] M. Ba, E. Pardoux, The effect of competition on the height and length of the forest of genealogical trees of a large population, in *Malliavin Calculus and Related Topics, a Festschrift in Honor of David Nualart*. Springer Proceedings in Mathematics and Statistics, vol. 34 (Springer, New York, 2012), pp. 445–467

[6] M. Ba, E. Pardoux, A.B. Sow, Binary trees, exploration processes, and an extended Ray–Knight Theorem. J. Appl. Probab. **49**, 210–225 (2012)

[7] J. Bertoin, J.F. Le Gall, The Bolthausen–Sznitman coalescent and the genealogy of continuous-state branching processes. Probab. Theory Relat. Fields **117**, 249–266 (2000)

[8] J. Bertoin, J. Fontbona, S. Martinez, On prolific individuals in a supercritical continuous state branching process. J. Appl. Probab. **45**, 714–726 (2008)

[9] P. Billingsley, *Probability and Measures*, 3rd edn. (Wiley, New York, 1995)

[10] P. Billingsley, *Convergence of Probability Measures* , 2nd edn. (Wiley, New York, 1999)

[11] L. Breiman, *Probability* (Addison-Wesley, Reading, 1968). New edition (SIAM, Philadelphia, 1992)

[12] P. Brémaud, *Point Processes and Queues: Martingale Dynamics* (Springer, New York, 1981)

[13] M.E. Caballero, A. Lambert, G. Uribe Bravo, Proof(s) of the Lamperti representation of continuous-state branching processes. Probab. Surv. **6**, 62–89 (2009)

[14] P. Cattiaux et al., Quasi-stationary distributions and diffusion models in population dynamics. Ann. Probab. **37**, 1926–1969 (2009)

© Springer International Publishing Switzerland 2016
É. Pardoux, *Probabilistic Models of Population Evolution*, Mathematical Biosciences Institute Lecture Series 1.6, DOI 10.1007/978-3-319-30328-4

[15] D. Dawson, Z. Li, Stochastic equations, flows and measure-valued processes. Ann. Probab. **40**, 812–857 (2012)

[16] J.F. Delmas, Height process for super-critical continuous state branching process. Markov Process. Relat. Fields **14**, 309–326 (2008)

[17] T. Duquesne, J.F. Le Gall, *Random trees, Lévy processes and spatial branching processes*, Astérisque **281**, Société de Mathématique de France (2012)

[18] S. Ethier, T. Kurtz, *Markov Processes, Characterization and Convergence* (Wiley, New York, 1986)

[19] J. Geiger, G. Kersting, Depth-first search of random trees and Poisson point processes, in *Classical and Modern Branching Processes*. The IMA Volumes in Mathematics and Its Applications, vol. 84 (Springer, New York, 1997), pp. 111–126

[20] A. Grimvall, On the convergence of sequence of branching processes. Ann. Probab. **2**, 1027–1045 (1974)

[21] A. Joffe, M. Metivier, Weak convergence of sequences of semimartingales with applications to applications to multi type breaching processes. Adv. Appl. Probab. **18**, 20–65 (1986)

[22] S. Karlin, H.M. Taylor, *A First Course in Stochastic Processes*, 2nd edn. (Academic, New York, 1975)

[23] A. Lambert, Population dynamics and random genealogies. Stoch. Models **24**, 45–163 (2008)

[24] A. Lambert, The contour of splitting tree is a Levy process. Ann. Probab. **38**, 348–395 (2010)

[25] J. Lamperti, Continuous state branching processes. Bull. Am. Math. Soc. **73**, 382–386 (1967)

[26] J.F. Le Gall, Marches aléatoires, mouvement brownien et processus de branchement, in *Séminaire de Probabilités XXIII*. Lecture Notes in Mathematics, vol. 1372 (Springer, Berlin, 1989), pp. 258–274

[27] J.F. Le Gall, *Spatial Branching Processes, Random Snakes and Partial Differential Equations*. Lectures in Mathematics ETH Zürich (Birkhäuser, Basel, 1999)

[28] V. Le, E. Pardoux, Height and the total mass of the forest of genealogical trees of a large population with general competition. ESAIM Probab. Stat. **19**, 172–193 (2015)

[29] V. Le, E. Pardoux, A. Wakolbinger, Trees under attack: a Ray-Knight representation of Feller's branching diffusion with logistic growth. Probab. Theory Relat. Fields **155**, 583–619 (2013)

[30] P.L. Lions, A.S. Sznitman, Stochastic differential equations with reflecting boundary conditions. Commun. Pure Appl. Math. **37**, 511–537 (1984)

[31] R. Lyons, Y. Peres, *Probability on Trees and Networks* (a book in progress), http://mypage.iu.edu/~rdlyons/prbtree/prbtree.html

[32] N. O'Connell, Yule process approximation for the skeleton of a branching process. J. Appl. Probab. **30**, 725–729 (1993)

[33] N. O'Connell, The genealogy of branching processes and the age of our most recent common ancestor. Adv. Appl. Probab. **27**, 418–442 (1995)

[34] E. Pardoux, *Markov Processes and Applications. Algorithms, Networks, Genome and Finance.* Wiley Series in Probability and Statistics (Wiley/Dunod, Chichester/Paris, 2008). Translated from the French original edition *Processus de Markov et applications* (Dunod, Paris, 2007)

[35] E. Pardoux, A. Rascanu, *Stochastic Differential Equations, Backward SDEs, Partial Differential Equations.* Stochastic Modelling and Applied Probability, vol. 69 (Springer, Cham, 2014)

[36] E. Pardoux, A. Wakolbinger, From Brownian motion with a local time drift to Feller's branching diffusion with logistic growth. Electron. Commun. Probab. **16**, 720–731 (2011)

[37] E. Pardoux, A. Wakolbinger, From exploration paths to mass excursions variations on a theme of Ray and Knight, in *Surveys in Stochastic Processes. Proceedings of 33rd SPA Conference* Berlin, 2009, ed. by J. Blath, P. Imkeller, S. Roelly. EMS Series of Congress Reports (2011), pp. 87–106

[38] E. Pardoux, A. Wakolbinger, A path-valued Markov process indexed by the ancestral mass. ALEA Lat. Am. J. Probab. Math. Stat. **12**, 193–212 (2015)

[39] J. Pitman, M. Winkel, Growth of the Brownian forest. Ann. Probab. **33**, 2188–2211 (2005)

[40] P. Protter, *Stochastic Integration and Differential Equations*, 2nd edn. (Springer, Heidelberg, 2005)

[41] D. Revuz, M. Yor, *Continuous Martingales and Brownian Motion*, 3rd edn. (Springer, Heidelberg, 1999)

[42] D.W. Stroock, S.R.S. Varadhan, Diffusion processes with boundary conditions. Commun. Pure Appl. Math. **24**, 147–225 (1971)

[43] J. Walsh, An introduction to stochastic partial differential equations, in *Ecole d'été de Probabilités de Saint–Flour XIV – 1984.* Lecture Notes in Mathematics, vol. 1180 (Springer, Heidelberg, 1980), pp. 265–430

[44] T. Yamada, S. Watanabe, On the uniqueness of solutions of stochastic differential equations. J. Math. Kyoto Univ. **11**, 155–167 (1971)

Index

© Springer International Publishing Switzerland 2016
É. Pardoux, *Probabilistic Models of Population Evolution*, Mathematical
Biosciences Institute Lecture Series 1.6, DOI 10.1007/978-3-319-30328-4

Printed in the United States
By Bookmasters